T0093522

Introducing Photonics

The essential guide for anyone wanting a quick introduction to the fundamental ideas underlying photonics, the science of modern optical systems. The author uses his 40 years of experience in photonics research and teaching to provide intuitive explanations of key concepts and demonstrates how these relate to the operation of photonic devices and systems. The reader can gain insight into the nature of light and the ways in which it interacts with materials and structures and learn how these basic ideas are applied in areas such as optical systems, 3D imaging and astronomy. Carefully designed end-of-chapter problems enable students to check their understanding. Hints are given at the end of the book and full solutions are available online. Mathematical treatments are kept as simple as possible, allowing the reader to grasp even the most complex of concepts. Clear, concise and accessible, this is the perfect guide for undergraduate students taking a first course in photonics and anyone in academia or industry wanting to review the fundamentals.

BRIAN CULSHAW is an Emeritus Professor of Optoelectronics at Strathclyde University, having previously served as Head of Department and Vice Dean. He is also a Director of OptoSci Ltd and was the 2007 President of the International Society for Optics and Photonics (SPIE).

"The field of photonics is the study of the generation, manipulation, and detection of electro-magnetic radiation (which we call light) whose energy is carried by what are known as photons. Professor Brian Culshaw has written a non-mathematical textbook that introduces the fundamentals of how light interacts with matter and reviews many of the diverse applications of optical systems in everyday life. This book is thus written at a level appropriate for undergraduate science or engineering students. Although advanced mathematical concepts are invoked (such as imaginary numbers and complex integration), mathematical derivations are avoided and simple explanations are used throughout the book. The author reviews the interactions between light and matter, including both absorption and emission of light, and then considers how the length scale of structures or features in matter can influence this interaction. Finally, the highlight of Professor Culshaw's book is a review of the many applications of photonics, ranging from lighting, digital communication, to even metal welding and cutting. For example, students are often surprised to learn that the internet and its functionality is enabled by the same form of energy that allows us to illuminate our homes after dark. Hence *Introducing Photonics* manages to successfully convey the diversity, excitement, and impact of photonic applications in modern society, without the mathematical details."

Kent D. Choquette, University of Illinois at Urbana-Champaign

"The high point of this inspiring little book is the 40-odd stimulating and thought-provoking questions at the end of each chapter."

Chris Dainty, National University of Ireland

"Brian Culshaw of Strathclyde University has been a long-time contributor to both photonics and ultrasonics. This book, Introducing Photonics, shares with the reader some of the knowledge about photonics the author has accumulated over his long career. The book contains much more on the optical properties of materials than many other books on this subject. I recommend the book to anyone who would like to benefit from the author's experience teaching photonics and performing cutting-edge research in this field."

Joseph Goodman, Stanford University

"The book that we were waiting for! The book that only a few very experienced people world-wide are capable to write! The book that we were waiting for to understand and/or clarify key photonic concepts in very easy, efficient and effective way! A key book for beginnings into the science and technology of light (photonics). It is an excellent tool for undergraduate students that aim to get a rapid, clear and effective view of photonic essential concepts, that also works for anyone to "refresh", efficiently, the fundamentals and technologies of light!"

Jose Miguel Lopez Higuera, University of Cantabria

"This book aims to introduce undergraduates to photonics and does it in a different and interesting format compared to typical textbooks. The author discusses many diverse topics in photonics together with achievements made possible by today's optical technology and what will be possible in the future photonics world. This book will encourage students to explore the ever-expanding field of photonics."

David A. Jackson, University of Kent

"Professor Brian Culshaw has been dealing with these issues for over 40 years, hence his knowledge transfer is extremely intuitive, deeply thought out not only from a scientific point of view but, above all, from the point of view of the recipient for whom photonics may be a new challenge. The minimal mathematical apparatus present in the book allows the reader to focus on an intuitive understanding of the discussed issue, thus creating an open path for him to search for new own applications. The concepts of ending each chapter with a set of problems for individual decision by the reader should be accepted as extremely valuable here. I think that their correct resolution is a challenge not only for the main recipients of the book - undergraduate students, but also for a wide range of active scientists involved in photonics. Hence, the book *Introducing Photonics* is an extremely valuable textbook appearing on the market."

Leszek R. Jaroszewicz, Military University of Technology

"*Introducing Photonics* is an enjoyable and thought provoking read for all with an interest in this field. This book is an excellent means to augment your existing knowledge and offers stimulating viewpoints on photonics from a long-standing expert in this area"

Kyriacos Kalli, Cyprus University of Technology

"An outstanding book that grasps the essence of photonics. Important concepts, from basic ideas to applications and future possibilities, are perfectly explained with easy language, clear descriptions and concise illustrations. University students in photonics as well as the more mature professionals in related fields will benefit from reading it."

Wei Jin, The Hong Kong Polytechnic University

"Professor Culshaw has written a must-read prime text for anyone considering entering the exciting field of modern-day photonics... all bases are covered... a comprehensive treatment from a fundamental introduction of the concepts behind light theory through to future photonic technologies and trends."

Elfed Lewis, University of Limerick

"This book provides a detailed explanation of the whole of the subject. It presents a new approach, concentrating on physical ideas, and developing them to explain more complex phenomena. This book is essential reading for practitioners, researchers, or students of physics and engineering whose work involves any of the large number of application areas touched by *Introducing Photonics*."

Yanbiao Liao, Tsinghua University

"The refreshing, intuitive style of this book makes it an excellent first text on light-based science and technology. Without sacrificing rigor, *Introducing Photonics* makes the field easily approachable; students with little more than basic science training will find this an enlightening and enjoyable text. Thought-provoking exercises challenge one to solve complete problems, rather than just use formulas to arrive at numerical answers by rote. In the final chapters, a sweeping panorama of current and future applications and innovations inspires the reader with the excitement and promise of the burgeoning field covered by this book: photonics, 'the electronics of the 21st Century'."

R. A. Lieberman, Lumopix, LLC

"We are just at the beginning of the age of light, the most sustainable and challenging age ever. Nothing like photonics has already changed and is still going to change for centuries our way of living, interacting, communicating, moving, caring and feeding ourselves. This book is a valuable and timely guide to review the key concepts of photonics and related applications through intuitive explanations and straightforward problems to check the understanding. From the basics of light-matter interaction, the reader is guided through consolidated photonic tools to the most advanced and intriguing devices, envisioning that the best of photonics has yet to come."

Anna Grazia Mignani, National Research Council (CNR), Italy

"Drawing on his vast experience in photonics research and innovation, Brian Culshaw presents the fundamental principles of photonics in an accessible manner, with end-of-chapter problems to reinforce the learning objectives. This is certain to be the go-to book for all students starting out in the field of photonics."

Sinéad O'Keeffe, University of Limerick

"*Introducing Photonics* is a surprising book! The profound and appealing way in which fundamental concepts are presented and what follows from them on applications in a multitude of fields of optics and photonics is remarkable, as well as the elegance and proximity with which the text is written.

At the end of each chapter a set of enlightening problems is proposed that follow the same framework, that is, its purpose is to induce thought and reflection on a particular phenomenon in the field of optics and photonics, as well as possible technological applications that may result from it.

The book is also prospective in a crystalline way regarding the future of photonics in its multiple areas, revealing from his author a deep knowledge and insight of the field.

In summary, I consider *Introducing Photonics* an inestimable book to guide and inspire a generation of students and practitioners in the field of optics/photonics."

José Luís Santos, University of Porto

"Brian Culshaw, in *Introducing Photonics*, succeeds in presenting a wide range of important topics in a clear, concise and effective manner that will be welcomed by students and technical workers. The linkages to real world applications throughout the text, problems and recommended readings serve as an effective entry point into this important field that has changed the world."

Eric Udd, Columbia Gorge Research, LLC

Introducing Photonics

BRIAN CULSHAW
University of Strathclyde

CAMBRIDGE
UNIVERSITY PRESS

University Printing House, Cambridge CB2 8BS, United Kingdom

One Liberty Plaza, 20th Floor, New York, NY 10006, USA

477 Williamstown Road, Port Melbourne, VIC 3207, Australia

314–321, 3rd Floor, Plot 3, Splendor Forum, Jasola District Centre,
New Delhi – 110025, India

79 Anson Road, #06–04/06, Singapore 079906

Cambridge University Press is part of the University of Cambridge.

It furthers the University's mission by disseminating knowledge in the pursuit of
education, learning, and research at the highest international levels of excellence.

www.cambridge.org
Information on this title: www.cambridge.org/9781107155732
DOI: 10.1017/9781316659182

First published 2020

Printed in the United Kingdom by TJ International Ltd, Padstow, Cornwall

A catalogue record for this publication is available from the British Library.

Library of Congress Cataloging-in-Publication Data
Names: Culshaw, B., author.
Title: Introducing photonics / Brian Culshaw.
Description: Cambridge ; New York, NY : Cambridge University Press, 2020. |
Includes bibliographical references and index.
Identifiers: LCCN 2019036970 (print) | LCCN 2019036971 (ebook) | ISBN 9781107155732
(hardback) | ISBN 9781316609415 (paperback) | ISBN 9781316659182 (epub)
Subjects: LCSH: Photonics.
Classification: LCC TA1520 .C85 2020 (print) | LCC TA1520 (ebook) |
DDC 621.36/5–dc23
LC record available at https://lccn.loc.gov/2019036970
LC ebook record available at https://lccn.loc.gov/2019036971

ISBN 978-1-107-15573-2 Hardback
ISBN 978-1-316-60941-5 Paperback

Additional resources for this publication at www.cambridge.org/culshaw.

Contents

Preface

Light has immediate and obvious appeal – the vast majority of us see light and communicate through it daily. We process images – facial expressions and traffic signs. We know light has an impact on people and places, emotionally and physically. Different colours affect our moods, sunshine lifts the spirit but with side effects for the excessively indulgent.

The word 'photonics' emerged into the scientific and technical vocabulary towards the end of the twentieth century and has come to mean much more than simply visible light extending into the redder than red and the bluer than blue.

Photonics is the science of light. It is the technology of generating, controlling, and detecting lightwaves and photons, which are particles of light. The characteristics of the waves and photons can be used to explore the universe, cure diseases and even to solve crimes. Scientists have been studying light for hundreds of years. The colours of the rainbow are only a small part of the entire lightwave range called the electromagnetic spectrum. Photonics explores a wide variety of wavelengths, from gamma rays to radio including x-rays, ultraviolet, and infrared light.

Photonics is also gaining much respect within the technical community; indeed 2015 was the International Year of Light and from 2018 onwards the International Day of Light will be marked on 21 May. But what makes photonics special? Indeed, what is it and how does it contribute to our emerging world?

These questions are easy to answer. Fibre optic communications are critical to the success of the internet, the biomedical community uses ever more versatile visible imaging systems exploiting subtle photonic concepts and phototherapy is now routine. Mobile displays and digital cameras are ubiquitous and there is much, much more to come.

The aim of this book is primarily to provide answers to the above two questions in concise and accessible manner. Yes, there are one-thousand-page tomes full of mathematics which expound every detail of the scientific principle. However, here we seek to highlight basic concepts and ideas and, most of all, attempt to develop some intuitive insight into what is going on.

Such insights, whilst not necessarily always totally accurate, are essential to understanding and formulating ideas and making that essential first guess (rough estimate) as to whether an idea is indeed feasible.

Some readers will be intrigued by photonics for its fascination alone as a technical discipline. For many the thought that this fascination may evolve into a diversity of application sectors over the coming decades will stimulate further curiosity and encourage broader participation. The final couple of chapters look very briefly into the diversity of prospects, both in applications and in scientific advances, to trigger yet more application potential. One aspiration here is to motivate yet further interest in this intriguing domain.

The book also includes problems at the end of each chapter, and the principal aim of these is to encourage readers to dig further and discuss the possible solutions among themselves. Indeed, many may respond best to teamwork approaches where different readers approach different pre-agreed tasks and then come together to evaluate a complete solution. Many of the problems are somewhat like 'real engineering' in that there is no perfect answer but the optimum approach needs to be fully understood and critically assessed by the readers. There will be discussion spaces on the web available as the book progresses.

For those wishing for more detail there are also many suggestions for further reading; there is a great deal of good material available on the web. Indeed, perhaps the real hope is that this book will stimulate the reader's curiosity towards discovering more about this fascinating topic.

Approaching and hopefully consolidating these insights has, from the author's perspective, emerged over many years with enormous thanks to countless encounters with groups ranging from students to internationally recognised experts; hopefully this book will consolidate at least the gist of these many experiences. Any list of these encounters is too long to mention and any attempt will inevitably have omissions.

There are also other acknowledgments to make, notably to the ever tolerant Sarah Strange at CUP whose constructively critical and wise comments on the manuscript have been remarkably helpful and thought provoking. All errors and omission, though, are purely the author's fault and comments from readers will always be welcome. Thanks too to Elizabeth Horne and Julie Lancashire for constructive support and understanding when yet another delay in the delivery of the manuscript appeared. Then there was Susan Parkinson, a splendidly understanding copy editor. Last, but by no means least, a very big thank you to the long suffering Aileen, who not only put up with the author in the household but also turned the manuscript from unintelligible ramblings into (we hope) a comprehensible text. We shall certainly celebrate when the book emerges.

Brian Culshaw

Glasgow

1
Photonics – An Introduction

Photonics is probably the outstanding candidate for the principal enabling technology of the twenty-first century. The term emerged around half a century ago and is now slowly creeping into everyday vocabulary just as 'electronics' did in the twentieth century. The aim in this short book is to introduce the principal concepts and ideas which underpin photonics and to explore its current and evolving technological significance.

1.1 What is Photonics?

'Photonics is the science of light. It is the technology of generating, controlling and detecting lightwaves and photons, which are particles of light. The characteristics of the waves and photons can be used to explore the universe, cure diseases and even to solve crimes. Scientists have been studying light for hundreds of years. The colours of the rainbow are only a small part of the entire lightwave range called the electromagnetic spectrum. Photonics explores a wider variety of wavelengths, from gamma rays to radio including X-rays, ultraviolet and infrared light.' This succinct definition comes from the website of the International Year of Light (IYL) celebrated in 2015.

The concepts through which we explore and understand light are indeed common throughout the electromagnetic wave spectrum, as indicated in the IYL definition. However, it is straightforward to appreciate that the scale of the wavelengths across this spectrum, extending from hundreds of metres for radio waves to subnanometres at X-ray frequencies, implies that different features within this common set will become more or less dominant in the understanding and application of a particular part of the spectrum. The transition from 'electronics' to 'photonics' reflects this gradual transfer, as indicated in Figure 1.1.

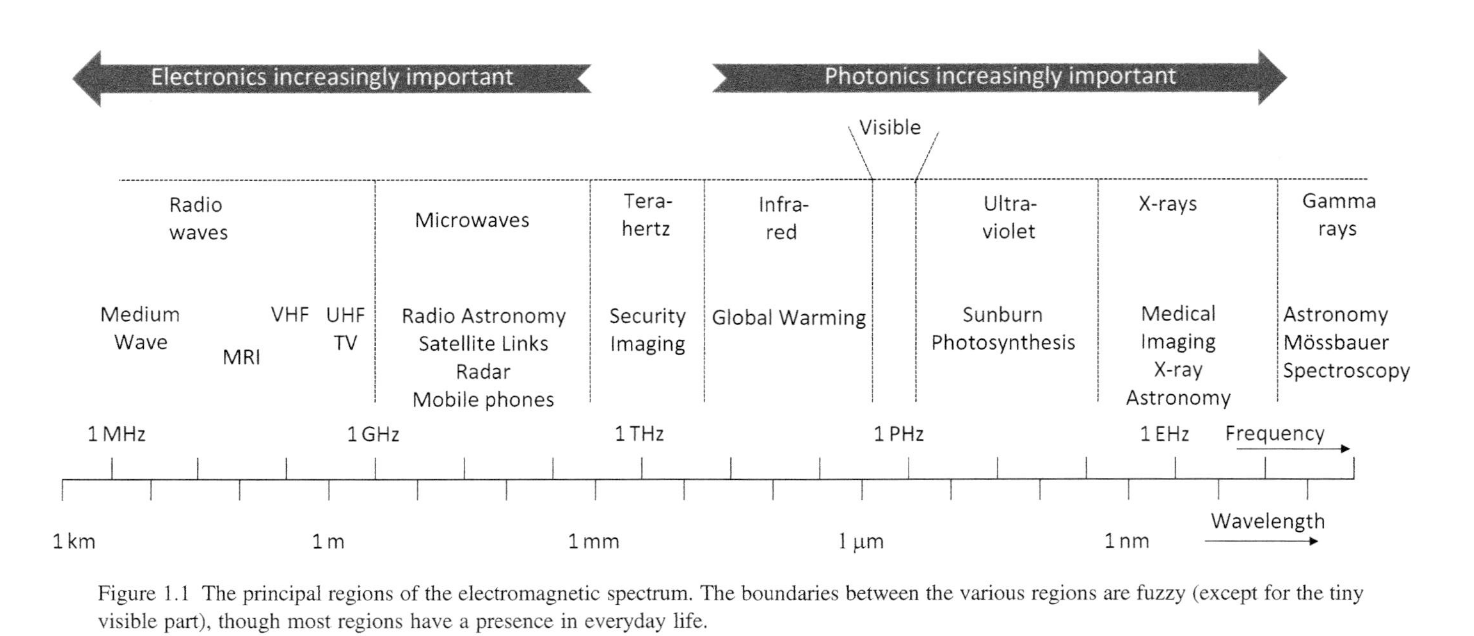

Figure 1.1 The principal regions of the electromagnetic spectrum. The boundaries between the various regions are fuzzy (except for the tiny visible part), though most regions have a presence in everyday life.

The focus in what follows is on those concepts which are most important in and around the visible spectrum, from the far infrared to the far ultraviolet. This has emerged as the more conventional domain for the term 'photonics'. The aspiration in this book is to convey an intuitive understanding of the essential photonics concepts and to develop a critical insight into applying them in order to appreciate the potential and, indeed, the limitations of new and emerging photonic technologies.

1.2 Exploring Some Concepts

The basic ideas of electronics and of photonics concern the interactions between electromagnetic radiation and materials. These interactions in turn depend fundamentally on both the material itself and also its dimensions compared with the wavelength of the electromagnetic radiation (Figure 1.1). This is exemplified through the observation that television is broadcast on radio waves with ultra-high frequencies – a few hundred MHz – where the wavelength is around the dimensions of a domestic antenna. From the antenna the signal proceeds into an electronic circuit – in your TV – which 'holds' the arriving electromagnetic wave within its much smaller dimensions. There is also an intermediate phase, in which the electromagnetic wave captured by your TV antenna is trapped within a cable before arriving at the electronic circuit. The flow of electrons in materials is common to the whole process, since the transmitted TV wave is generated from electrons flowing through a suitable transforming device to produce a propagating electromagnetic wave.

The source of the term 'photonics' lies in the concept of the photon. Photons, discrete quanta of light, were inferred through the 'ultraviolet catastrophe' in the black body spectrum and later more clearly defined through Einstein's description of the photoelectric effect, in which electrons are emitted from a material only if sufficient energy is acquired from the incident light. The required energy depends not on the incident power but on the optical frequency. This is in contrast with what is happening in a TV antenna which modifies the movement of already free electrons in a way that is directly dependent on the total power. However, it was already known that in other structures, lightwaves behaved in exactly the same way as radio waves, for example in transmitting through the atmosphere, being reflected at dielectric boundaries or interfering with themselves after transmission through different paths.

In photonics the threshold of 'where photonics matters' is now associated with the energy of the photon. The term 'photonics' becomes applicable when the photon energy becomes comparable with or exceeds the thermal energy of the particles in the material with which the electromagnetic wave is interacting. The energy of the photon, namely Planck's constant multiplied by the frequency of the electromagnetic wave, and the energy of a freely moving particle, which is of the order of Boltzmann's constant times the absolute temperature, are compared in Figure 1.2 at a temperature of 300 K. (The photon energy here is presented in electron volts – the energy change as an electron changes its electrical potential by 1 V – see Appendix 5 for the values of these constants.) When the photon energy comfortably exceeds the inherent thermal background energy of particles (at a wavelength of about 30 microns at 300 K – the far infrared), then electromagnetic radiation is captured or, indeed, generated as a particle. At lower energies (wavelengths above about 300 microns, 1 THz) we can regard the capture or generation of electromagnetic radiation as being due to a flow of electrons produced by the electric field in an electromagnetic wave. In other words, at lower frequencies a wave model for the interaction becomes appropriate; the ideas of 'electronics' rather than 'photonics' apply. This will depend on temperature: at 3 K a photonic description of an interaction will begin to be applicable at frequencies two decades less than at 300 K.

Similar observations apply to structural dimensions, though the boundaries between the photonic and electronic descriptions are based on different criteria. Materials behave 'electronically' when their dimensions are significantly less

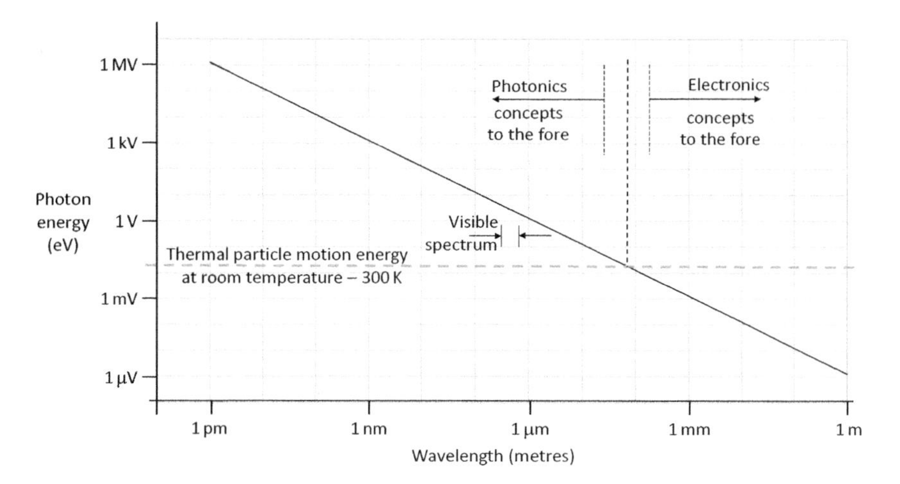

Figure 1.2 The plot shows photon energy as a function of wavelength. The broken line shows the energy of thermal random motion at 300 K. At wavelengths below about 30 microns (10 THz), the photon energy begins to dominate. At about 300 microns (~1 THz) the ideas of electronics become more applicable. The THz region falls in the transition zone.

than a wavelength. A contrasting aspect of photonics has been that very small scale structures with dimensions in the photonics region have only recently become potentially achievable; that is, to dimensional precision measured in nanometres. It is remarkable that gold nanospheres have been used to provide a spectrum of permanent colour in glassware since ancient times. Finally, there is the dimensional range of structures which is comparable with the wavelength and it is here that diffraction effects, in particular, become critically important and waveguiding – as in fibre optics – comes to the fore.

These principles and their context in potential applications will be explored in more detail later in the book. They have, however, prompted considerable debate as to whether light is 'really' a wave or a particle. The answer to this question is that it depends on the context. Our aspiration here is to help develop an understanding of which approach is relevant and when. An attempt to portray these differing situations is presented in Figure 1.3, which highlights the underlying phenomena that constitute photonics.

1.3 Applications

Applications, achieved and potentially feasible, are becoming a major stimulus for the interest in and development of photonics. We shall explore briefly a number of case studies later in the book, but among the everyday presence of photonics sits the internet, which could not function at all without fibre optic communications. Your camera, your smart phone and the increasing efficiency and versatility of lighting technologies all rely on recent advances in photonics. Many aspects of medical diagnosis, microscopy and medical therapeutics, rely on photonics. Optical absorption is the initiator for climate change; the monitoring processes measuring changes in our atmosphere, as well as the long-term solutions aimed at driving our energy sources directly from light, all need an understanding of photonics and its exploitation.

1.4 About This Book

The essential concepts which are covered in the book are shown in Figure 1.4. The next three chapters are devoted to the principles of the subject. Having established the principles, we shall explore some current applications through brief case studies. Finally comes a speculative look at the future in terms of both the enabling technologies and the technological advantages, together with some social needs to which photonics may contribute to in the future either

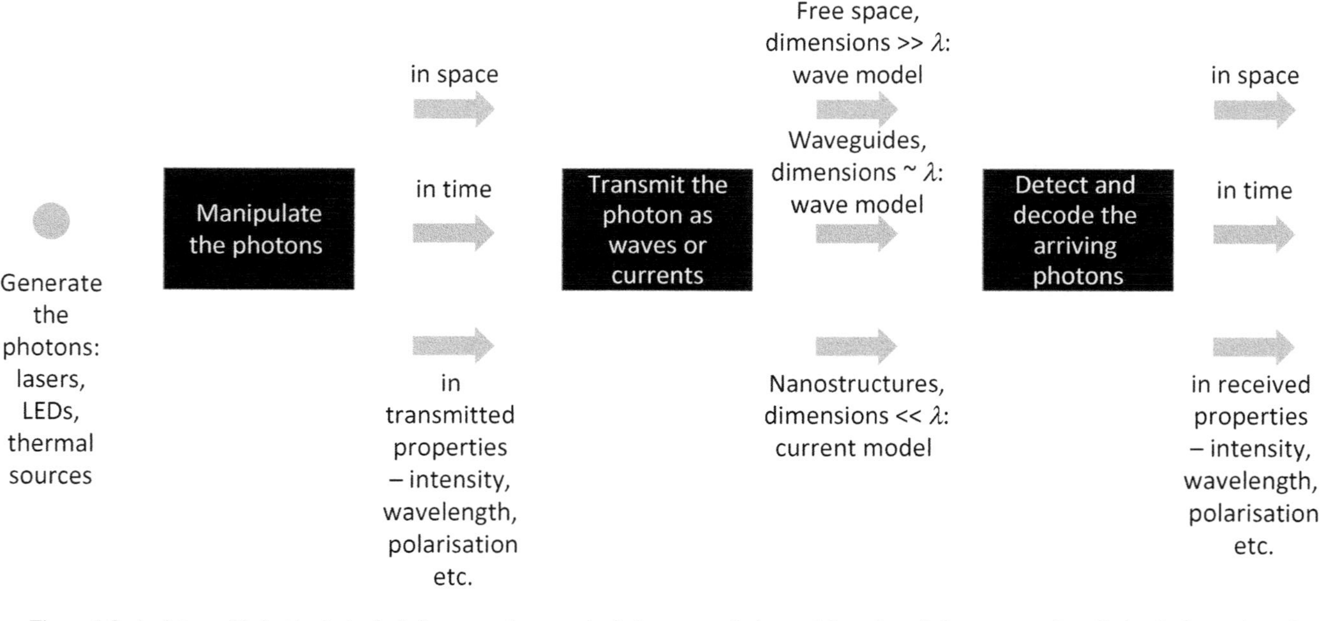

Figure 1.3 A picture of 'what is photonics': the generation, manipulation, transmission and detection of electromagnetic radiation in that region of the spectrum where the photon picture dominates the behaviour at generation and detection, but where transmission can be described in terms of waves, in large-scale dielectrics or in medium-scale waveguides. At the smallest scale, over very short distances, currents in conductors or in nano-scale synthetic dielectric structures, become important.

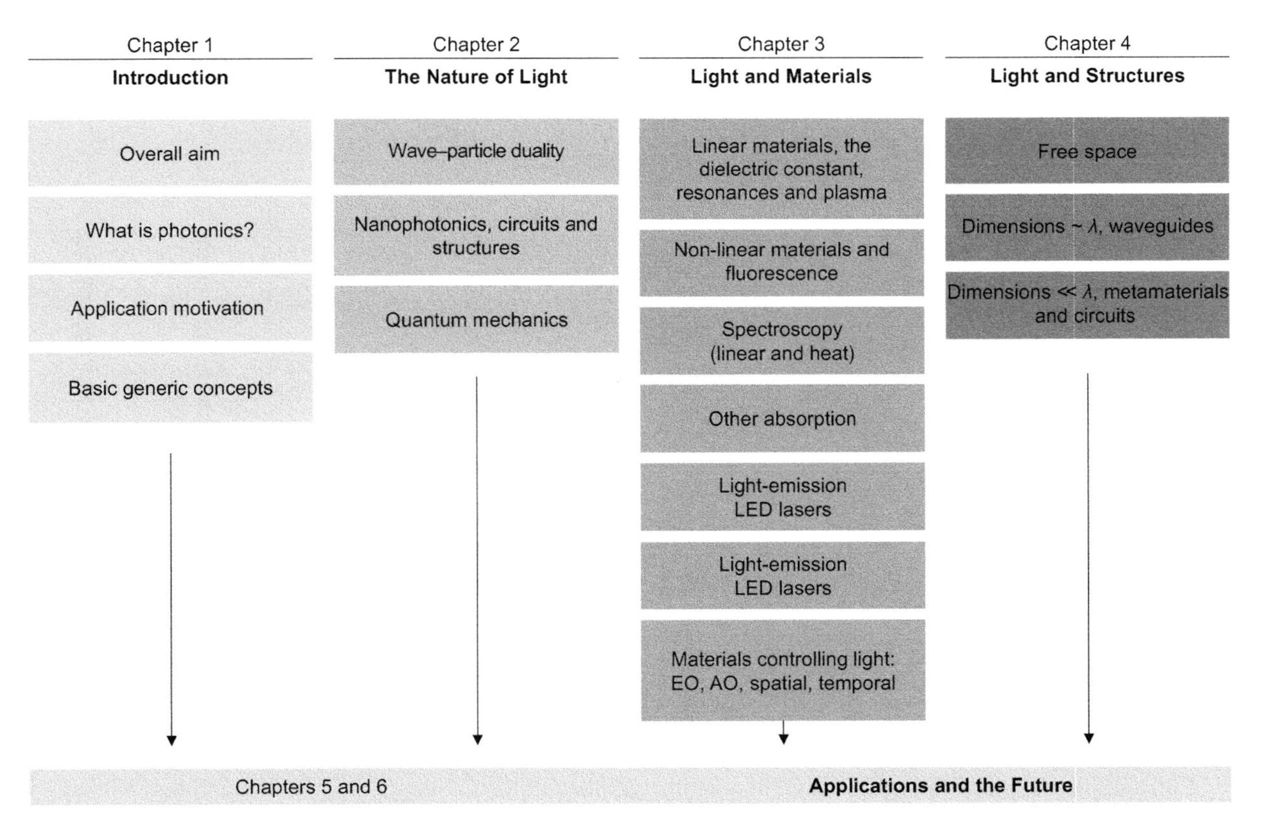

Figure 1.4 The essential concepts of photonics and their organisation in this book.

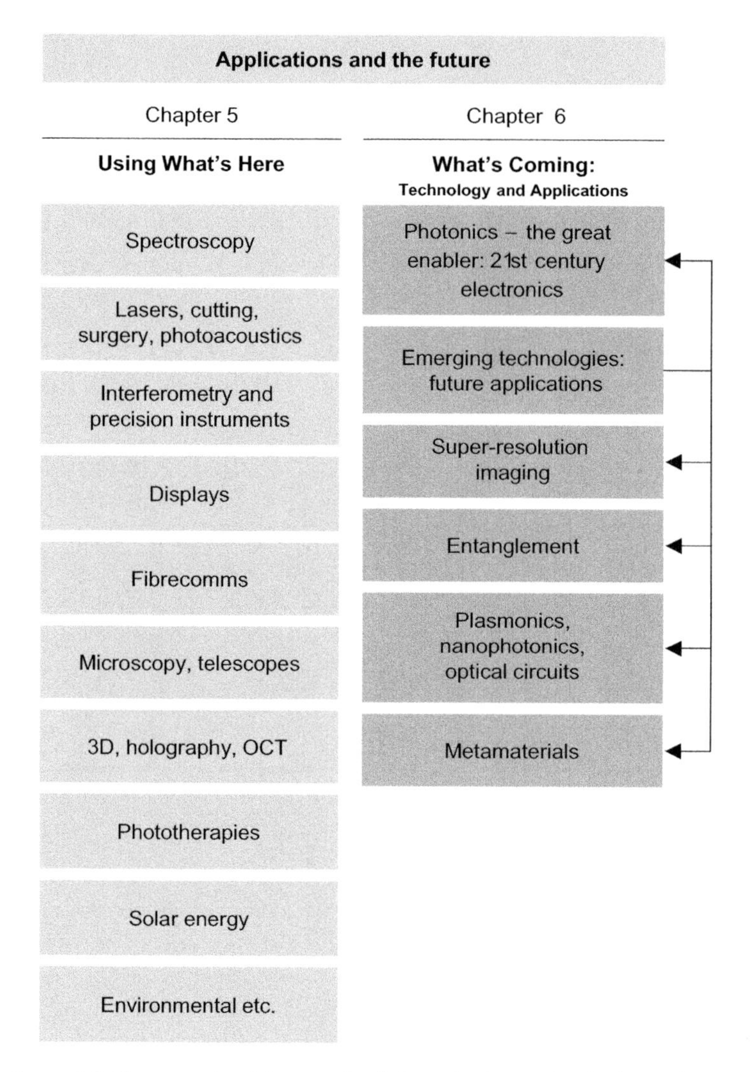

Figure 1.5 The present and future role of photonics in our developing society.

through further exploitation of what we already know or through the use of the new tools, as encapsulated in Figure 1.5.

1.5 Some Related Background Ideas

The best way of describing the interaction between light and a material depends on the nature and structure of the material and its dimensions. Resonance plays a large part in photonics; this occurs when the structural dimensions match the wavelength. The same phenomenon is familiar in musical instruments; long organ pipes resonate at lower frequencies than short ones. Resonant structures have entirely different propagation properties from an unstructured equivalent space and this applies in every day optics too, from the coatings on camera lenses to the colours of butterfly wings. The phenomenon whereby the dimension of a structure, in relation to the wavelength of an incident wave, affects the material properties seen by the wave is ubiquitous and powerful; it applies to every form of wave from mechanical resonances and acoustic waves to visible light and into the X-ray region. The same concept also explains much of quantum mechanics, particularly the ideas of energy levels and band structure. The concept of the 'universal wave' will be used extensively in what follows and, for a deeper and more thorough examination of this very powerful intuitive tool, there is little if anything to match J. R. Pierce's book (Further Reading A). Other conceptual tools are also extremely useful for an understanding of photonics; of these perhaps the *Fourier transform* is the most important. There is an extensive literature available, including a classic text by R. Bracewell and one by J. F. James (both listed in Further Reading A). Much of what is discussed here is also underpinned by Maxwell's equations, which themselves describe many of the interactions between electromagnetic waves and materials. Daniel Fleisch has produced a straightforward guide to these important relationships (see Further Reading A). At the other extreme, whilst the present text aims to introduce the basic concepts of photonics, the book by Saleh and Teich (see Further Reading C) presents over 1000 pages of comprehensive detail for those who may wish to go further.

Finally, the International Year of Light produced some inspiring texts (see Further Reading B) highlighting the critical contributions which light makes to our world and exploring the many photonic tools we can use to aid this exploration ranging from a mirror through a smart phone to the Hubble telescope.

1.6 Problems

Some topics to investigate:

1. How might the boundary lines for photonics, as indicated in Figure 1.2, change in deep space? Could there be circumstances where, for example, the photon concept is useful for describing an electromagnetic wave of 3 cm wavelength?

2. The term 'microwave photonics' appears fairly often in current scientific literature. What is likely to be the prime condition for the observation of microwave photons? Figures 1.1 and 1.2 might give some hints on this. Note that microwave photonics can also apply to the situation where a light beam is modulated by a microwave frequency signal.
3. Now think about a microwave communications satellite, in synchronous orbit about 22,400 miles above the earth where the ambient temperature is a few kelvins. Would you consider a communications satellite a photonic device and why? Some points to consider are that it is subject to the sun's rays for half the day, the internal circuits generate heat and the heat loss in what is essentially a vacuum will be primarily by radiation. *(It may be helpful to look up the satellite communications frequency band using a search engine. A great deal of useful background on many, many topics can be found in this way!)*
4. At the other end of the spectrum – when might X-rays with photon energies in the 100 V to 100 kV range behave as waves rather than particles?

2

The Nature of Light

The question 'is light a ray, a wave or a particle?' has long fascinated philosophers and has intensified since the concept of the photon first emerged at the beginning of the twentieth century. The pragmatic answer to the question is 'it depends what you are looking for' and is really about how best to conceptualise light in particular circumstances. We have already mentioned such distinctions briefly in Chapter 1. This chapter will extend these basic concepts to introduce the material which follows in later chapters.

2.1 Light in a Vacuum

This is the simplest case. When no other materials impede its progress or modify its path (Figure 2.1), propagation in a vacuum can be accurately described through a wave propagation model. Ray optics also works since in an idealised infinite vacuum there are no edges or multiple paths through which the light can interfere or diffract.

By the same token, however, a particle of light will also proceed unimpaired along the same path as the ray. And, already, we have an apparent contradiction with the wave propagation model, in that a single photon cannot spread its presence over a surface area increasing as the square of the distance from the source. This dilemma continues to produce philosophical debate but really only becomes pertinent when we produce single identifiable photons and watch them in motion. More on this later!

Perhaps the principal useful aspect of propagation in a vacuum is that light then proceeds at the velocity commonly approximated as $c = 3 \times 10^8$ m/s.

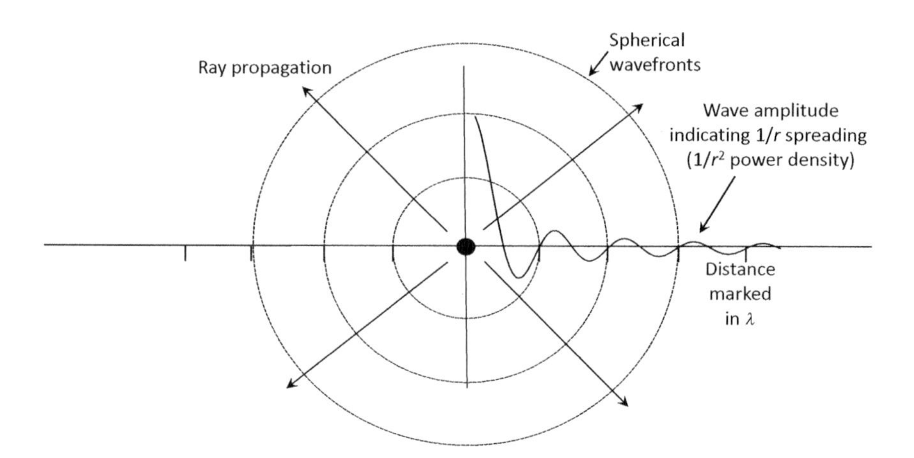

Figure 2.1 Propagation of rays and waves from a point source in a vacuum. The wave amplitude representation indicates a reduction in power density with the inverse square of the distance, but the total power remains constant as the wave progresses.

2.2 Light in Isotropic and Anisotropic Materials

In a material, light, as it propagates, interacts with the molecules within the material, typically by inducing some form of electronic motion within the molecular structure; this may result in dipole moments which change in response to the lightwave's electric field (Figure 2.2).

We can regard the way in which this dipole moment follows the incoming electric field as like a classical mass–spring–damper oscillator excited by the electromagnetic wave. This process (Figure 2.3) can be viewed in more detail through the Clausius–Mossotti equation, discussed in Appendix 3, to give a surprisingly useful and accurate insight into the behaviour of the material's dielectric constant ε (see Appendix 3) and, through that, its refractive index n as a function of frequency, $n = \sqrt{\varepsilon}$, is probably the most useful concept in optics and photonics and is defined by equation 2.1, where v is the velocity of light within the material:

$$n = \frac{c}{v} \tag{2.1}$$

The slowing down of light in a material can be viewed as due to the energy storage and release through the electron motion induced in the material, which introduces a delay in propagation. Through similar reasoning (and confirmed through the Clausius–Mossotti equation) the more energy that is stored the larger the delay, so the more molecules per unit volume the higher the

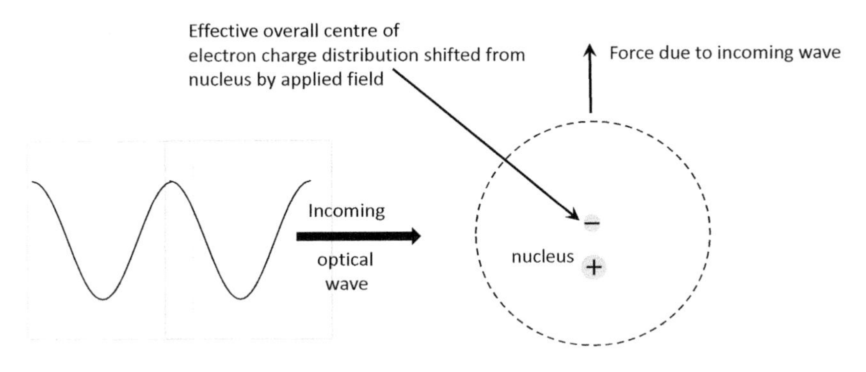

Figure 2.2 A schematic of how the oscillating electric field in an incident electromagnetic wave can distort the electronic charge distribution around an atom, producing an oscillating dipole.

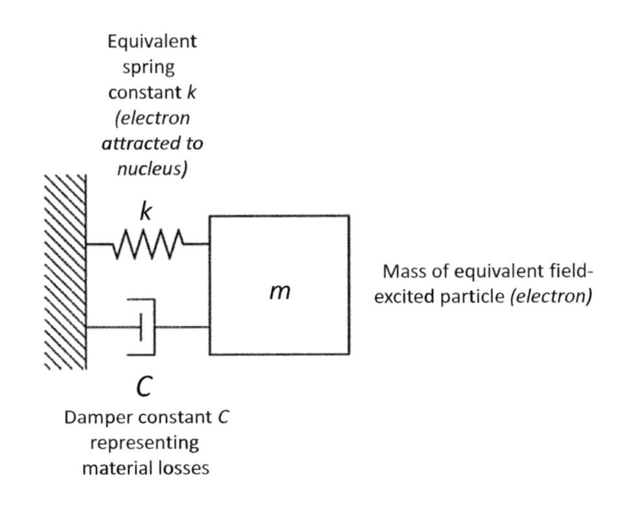

Figure 2.3 The mass–spring–damper model of an atomic electron forced into oscillations about the nucleus by an optical wave, thus forming an oscillating dipole.

refractive index. By the same token, in the case of anisotropic materials, exemplified in many crystalline structures, the refractive index seen by an incoming electromagnetic wave will be higher when its electric field oscillates in a crystalline plane with a higher density of molecules. An electric field oscillating in a plane with lower density will experience a lower refractive index and so will travel faster. (This phenomenon is known as *birefringence*.) Hence the different polarisation states (Figure 2.4) of an incident light beam with respect to the material's orientation will see a range of refractive indices.

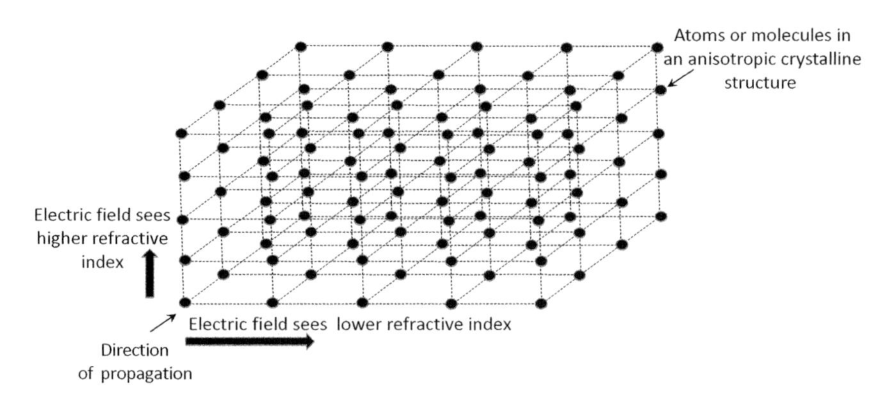

Figure 2.4 The origin of birefringence in asymmetrical materials – the effective molecular density perceived by the incoming wave depends upon its electrical polarisation direction.

Another important observation here is that the refractive index for polarisation states other than the two states aligned with the crystalline structure of the material will be a mix of two values – the input state will be split into these two components along the so-called *principal axes* of the material. Each component will consequently exit with a different phase delay, so that the output polarisation state will in general be different from the input state. Appendix 1 explores this in more detail.

There are many other insights which are enabled by the mass–spring–damper model. Any mass–spring–damper arrangement involves a frequency response to a given applied force. At low frequencies the deflection introduced by the applied force has a particular characteristic value and is in phase with the applied force. At the *resonant frequency*, the deflection is a maximum but is now in quadrature with the applied force. After passing through the resonance the displacement becomes small and in antiphase (Figure 2.5).

By the same token, then the refractive index will have exactly the same general resonance characteristics, as the mass–spring–damper model though in molecular materials there are many resonances and therefore the refractive index behaviour follows the trends indicated in Figure 2.6. Consequently, since all these resonance effects are taking place, the refractive index does vary with frequency and so the velocity of light within the material is also a function of frequency – a phenomenon known as dispersion. Looking at an everyday example, at low frequencies water has a dielectric constant of 80, corresponding to a refractive index of about 9, but the refractive index in the visible region is about 1.3.

Figure 2.6 also imples a number of other important photonic features, common to all materials. The refractive index generally decreases with

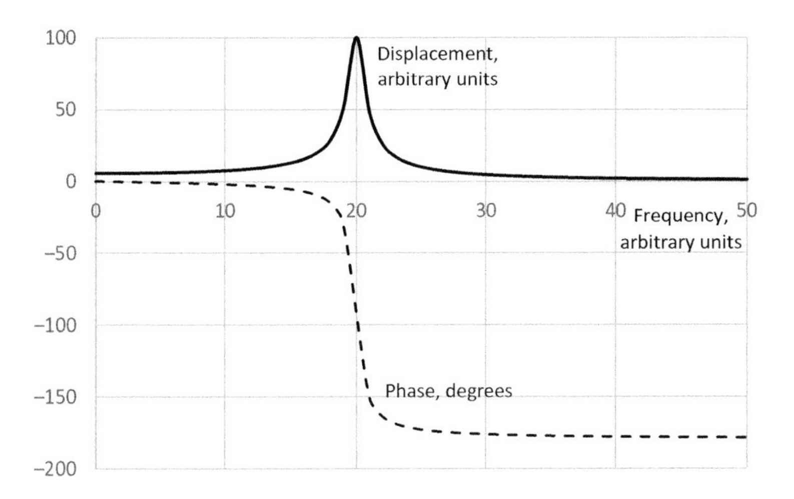

Figure 2.5 The generalised displacement and phase of a mass–spring–damper oscillator as functions of the frequency of the applied force. Note the change in phase around resonance and the fact that the displacement is in antiphase above the resonant frequency, resulting in a negative contribution to the dielectric constant.

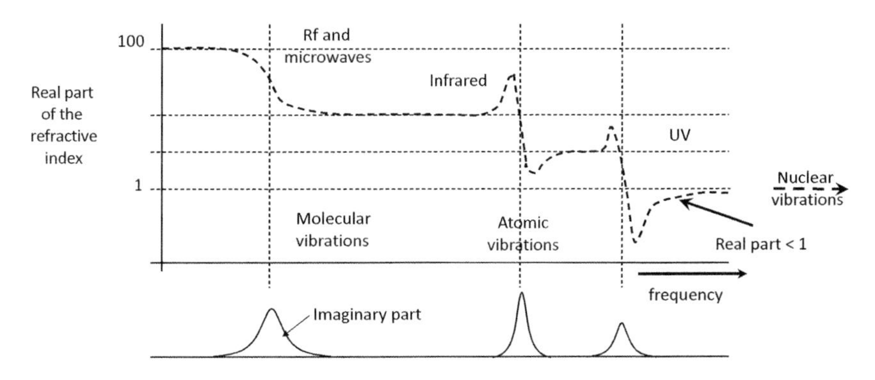

Figure 2.6 The generalised behaviour of the refractive index as a function of frequency indicating resonances due to various molecular and atomic vibrations.

frequency except around the resonances, after which there is another step downwards. Also, the refractive index can drop below unity. But does the velocity of light ever exceed its velocity in a vacuum? The implications of this apparent dilemma are examined briefly in Appendix 4, which considers the phase and group velocity concepts.

Thus far we have ignored another important aspect of our mass–spring–damper model – namely, the damper. This represents the sources of loss, and this lost energy is converted typically into heat. So, strictly speaking, the refractive index, which describes the transfer of energy through the material, should include this loss term. This is expressed through the representation $n' = n + jk$, where j is the square root of -1 and k is the known as the imaginary component of the refractive index. Additionally, as indicated in Figure 2.6, the loss terms increase significantly at resonance since this is where the radiation is more readily absorbed, typically by a large factor. This absorbance feature is one source of the colour in the world around us.

The propagation of light within a homogenous material is then in general described through the propagating wave equation

$$E(z,t) = E_0 \exp\left[-j\left(\omega t - \frac{2\pi nz}{\lambda}\right) - \frac{2\pi kz}{\lambda}\right] \tag{2.2}$$

Here E is the electric field amplitude of the propagating wave, z the axis along which the wave propagates, ω the optical angular frequency in radians/second and λ the wavelength in vacuum; and the final term represents the attenuation of the electromagnetic wave. The first term gives the time dependence, and the second term the distance dependence, of the optical phase. Note that the imaginary part k of the refractive index determines the level of attenuation.

There is yet another aspect to this – suppose the electrons are no longer bound to individual molecules but free to move. This happens, for example, in conducting metals and also in the ionosphere around the earth. In this case the restoring force is due to the electrostatic forces exerted by the positive ion lattice, as indicated schematically in Figure 2.7. Remember, too, in the case of

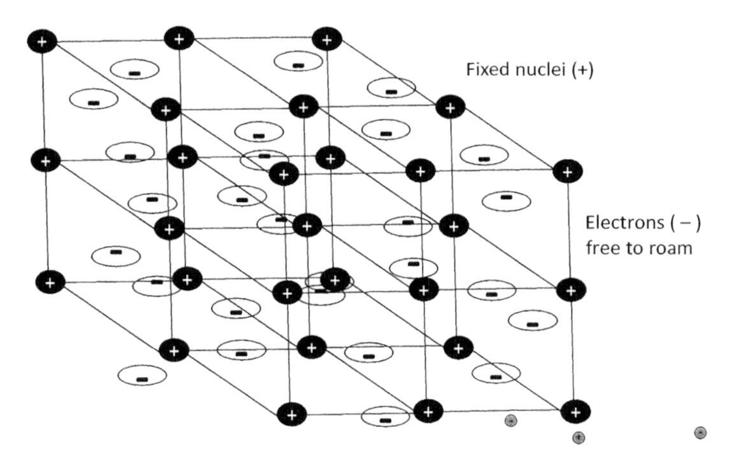

Figure 2.7 The plasma case where free electrons experience a restoring force from stationary positive ions. Note there is minimal damping for frequencies above the inverse of the electron lifetime in the material.

conductors, the losses are due to the collisions between the moving electrons and the stationary ions, a phenomenon encapsulated in the electron *relaxation time* for the material concerned.

There is again a moving mass, namely that of the electrons, and so again we have a resonance dependent upon the density of the free electron cloud and given by (see Appendix 3):

$$\omega_p^2 = q_e^2 N_0 / m_e \tag{2.3}$$

where q_e and m_e are the electronic charge and mass, N_0 is the free electron density per cubic metre and ω_p is known as the plasma resonance (angular) frequency.

This plasma resonance has many interesting ramifications. Everyone is aware that X-rays travel through metal but light does not. Satellite broadcasting in the GHz region comes down from outer space through the ionosphere. However, much lower-frequency radio waves in the short wave band bounce off the ionosphere in the same way that light bounces off metals.

The impact of the relaxation time in plasmas (see below) is also important. As mentioned above, the conduction losses in a metal are basically determined by the frequency of collisions between the freely flowing conduction electrons and the stationary ions in the metal. However, if we apply an electric field at a frequency which exceeds this collision frequency, the inverse of which is the *relaxation time*, then there are many fewer collisions between the conduction electrons and the stationary particles in the metal, so the losses decrease dramatically – in other words, the metal becomes transparent.

The refractive index of gold as a function of wavelength is indicated in Figure 2.8. Among other things the rapid variation of the complex index with wavelength in the visible region (roughly 0.4–0.8 μm) accounts for the difference in the colour of gold as compared, for example, to the colour of silver. Most metals in and around the optical and ultraviolet region exhibit an intriguing transition between relatively low-frequency behaviour as a conductor and the higher-frequency behaviour as a relatively transparent dielectric. In other words, the electrons in the metal constantly turn around so that there is no current flow and likewise no collisions to cause absorption losses. This fascinating transition has, within the past couple of decades, evolved into so-called *plasmonics*, which seeks to understand and exploit this somewhat complex evolution.

This discussion has been about light as a wave. There has been no mention of the photon. These classical wave approaches do give a very reliable guide and an indispensable insight into the interaction between light and materials. The photon model also has its place, however. This is most apparent with

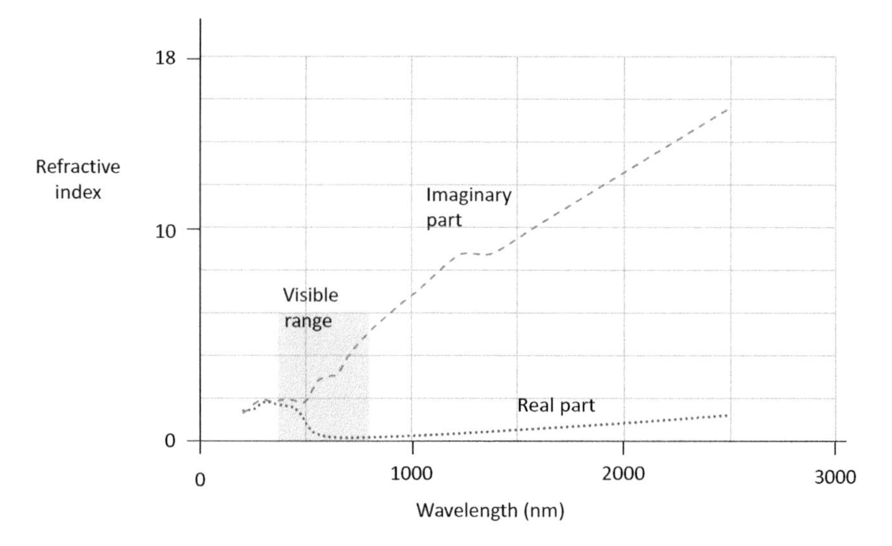

Figure 2.8 The refractive index of gold indicating its change in behaviour emerging at frequencies above the plasma resonance, corresponding to about 200 nm wavelength.

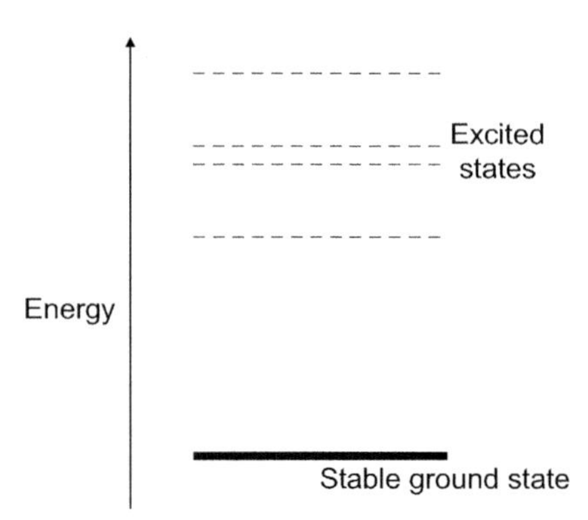

Figure 2.9 An energy level representation for a single atom – which approximates to the behaviour of a gas at low pressure.

regard to the energy levels evident in a material (Figure 2.9), a concept which has evolved as the 'conventional' approach to explaining the wavelength-selective absorption of light in materials.

The differences between these energy levels and the ground state in a molecule equal the photon energy required to excite the molecule from the stable ground state. This absorbed energy then typically (but not always) re-emerges as heat when the molecule relaxes to its ground state. The same behaviour can be ascribed to the resonant absorption of mass–spring–damper systems if we are looking purely to match the frequency of the input light with the frequency of the mechanical resonance. This merging of the wave (and mechanical resonance) domain and the photon domain (via energy level differences) can provide useful insights into material behaviour.

There are lower-frequency transitions in molecules as well, corresponding to, for example, exciting an entire molecule into rotational rather than electron-orbit resonances (see Figure 2.6 for the general trends). A microwave cooker which operates at around 2.5 GHz exploits exactly this effect, tuning to resonances in water, fats and sugars. Here, however, we are by no means concerned with a photon energy in the order of a few microvolts; this is well below the photonics region discussed in Chapter 1. The scientific community also talks of these frequencies in terms of molecular resonances, implying an analogous electromagnetic concept.

Much of the wave and photon dilemma is encapsulated in the above discussion. Whilst numerous weighty texts have been published on the subject there is really no definitive answer. For us it amounts to finding the approach which gives the most useful model to analyse and understand the photonics scenario of interest.

2.3 Light Interacting with Structures

A material structure, in our present context, is an arrangement which comprises two or more materials with different optical properties. Once again, the question of a structure's scale compared with the optical wavelength enters the discussion.

So, we start with reflection and refraction at a large-scale structure, namely an interface between two dielectric materials (Figure 2.10). We will simplify the discussion by assuming that the interface is planar. The light that passes through the interface is refracted by an amount determined by *Snell's law*, namely:

$$\frac{n_r}{n_i} = \frac{\sin \theta_i}{\sin \theta_r} \tag{2.4}$$

The reflected light obeys the normal 'angle of incidence equals angle of reflection' relationship, but what is interesting here is the amount which is reflected and the way in which this amount depends upon the incident angle. Indeed, at one specific angle – the *Brewster angle* – only one polarisation state, namely that with the electric vector aligned with the surface of the interface (sometimes called S (in-plane) polarisation and sometimes called TE for transverse electric polarisation) is reflected. The reflection coefficient is not 100%, so that the refracted ray still contains some of this polarisation state. The reflection coefficient as a function of angle for the two linear polarisation states

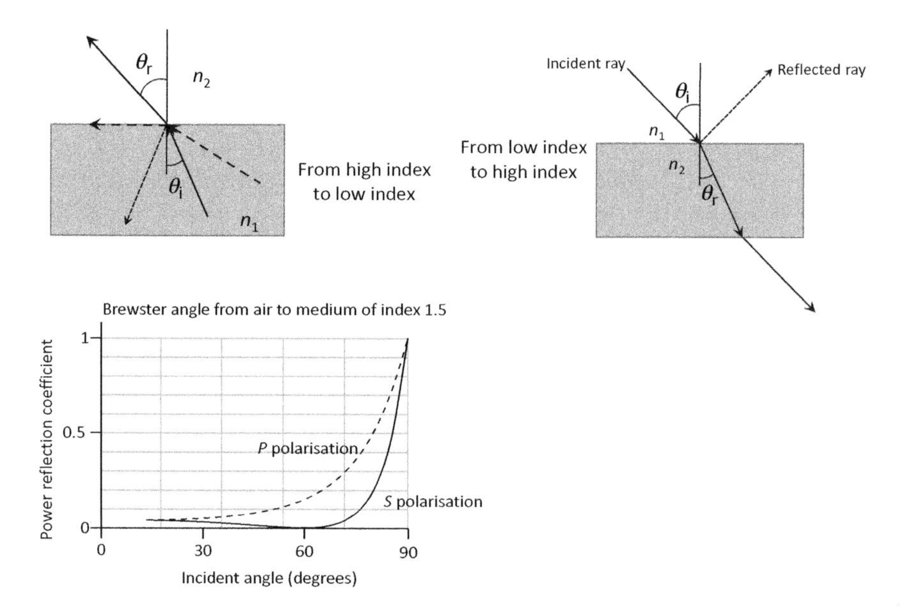

Figure 2.10 Reflection and refraction at a planar interface between two materials differing in refractive index. The diagram at top left shows a ray striking the interface at angle θ_i and being partly transmitted at angle θ_r and partly reflected back into the higher-index material. The ray shown with a broken line is incident at the critical angle, and so would travel along the interface. At incident angles greater than thus, total internal reflection occurs. The Brewster angle plot applies to an index of 1.5 in the higher-index material for light incident from the low-index side of the interface.

is also shown in Figure 2.10. The partial removal of the S polarisation state (with electric field vibrations perpendicular to the P state vibrations and to the direction of the reflected ray) manifests itself in daily life, amongst other things in the polarising sunglasses designed to reject the (partially) polarised reflected light from our surroundings.

The discussion thus far concerns a light beam travelling from a low-index medium to a higher-index medium. When the situation is reversed, there will be a case when the angle of incidence is high enough for refraction to occur into the interface plane (i.e. the angle of refraction is 90°), and then the incident angle is referred to as the critical angle. If the incident angle goes beyond that, total internal reflection occurs, as implied in Figure 2.10, which also indicates some reflection for incident angles below the critical angle.

Incidentally, the derivation of Snell's law (see the Chapter 2 problems) exemplifies a very important principle in understanding the coupling of light from one material or structure into another. This principle is that the projected components of the wavelengths along the interface must be the same – a concept known as phase matching. This idea applies to any wave propagating through an interface, whether the wave is acoustic, electromagnetic or even a water wave.

There are many other possible interfaces. In the above we have assumed a perfectly planar interface. However, slight roughness will introduce corresponding slight changes in the angles of incidence and refraction until, in the limit, a perfectly scattering surface would send the light in all directions. Again we encounter this daily – sunlight entering your room is scattered throughout the entire volume of the room, by the walls and furniture.

Moving on to a medium-scale structure reaches into diffraction and interference: we move from the ray optic approach, which works at large scales, into the wave optics approach. Much is encapsulated in the classic Young's slits experiment, shown in Figure 2.11. Light from a single-parallel-beam single-wavelength source impinges on two slits, separated by a distance d and each of a width conveniently assumed to be much less than the wavelength. The input light then reradiates in all directions and produces an interference pattern as indicated. The interference pattern is determined by the vector sum of the fields directed through the two paths. Full constructive interference occurs when the path difference $d_A - d_B$ is an integer number n of wavelengths, which gives, for $y \gg d$,

$$n\lambda \approx \frac{dx}{y} \tag{2.5}$$

where x is the distance on the screen from the axis of the system and y is the distance along the axis between the slits and the screen. There is a time delay

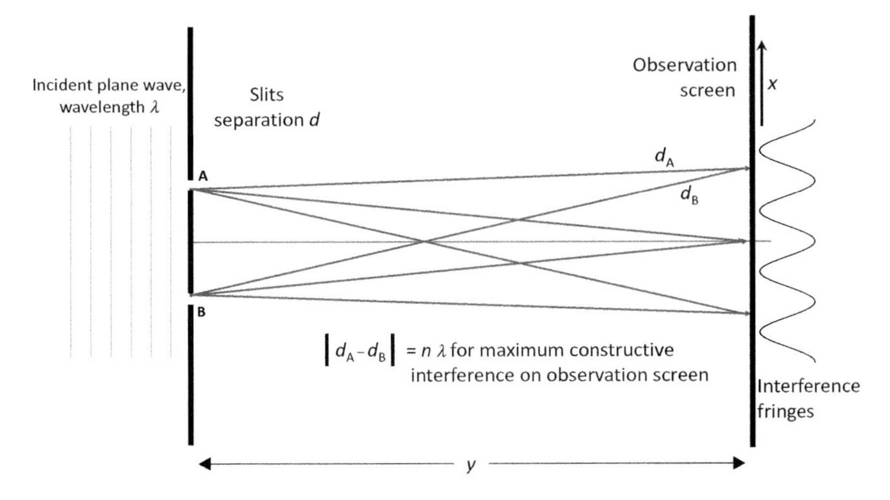

Figure 2.11 Young's slits – the simplest geometry to exhibit interference phenomena.

between arrivals at a particular point on the screen from the two paths and there is an implicit assumption in equation 2.5 that the phase delay on the optical signal is completely predictable for this time difference. Sometimes it is not, which leads us into the concept of coherence; ultimately this coherence concept applies to both the spatial and temporal behaviour within an optical source, a subject to which we shall return later.

Much of photonics concerns this medium-scale structure, ranging from camera lenses and image sensing arrays to DVD players and the display on your telephone handset. Furthermore, almost all of this discussion is concerned with wave optics.

At small scales we move into the domain of subwavelength structure in a material. This domain has only recently become available, with the notable exception of materials such as colloidal gold, which has been used in pottery for centuries and for which the colour can range through reds to blue depending upon the diameter of the tiny gold spheres within the colloid. The resonance wavelength for the electron plasma circulating the spheres varies as indicated in Figure 2.12, but gold spheres throughout this dimensional range also absorb more in the blue than the red, owing to material absorption, the property which makes 'normal' gold appear golden. In current terminology these gold spheres would be regarded as an example of quantum dots. As colloidal gold they were simply a way of introducing permanent colour into artefacts. The tiny structural dimensions take the 'normal' colour far from its golden origins.

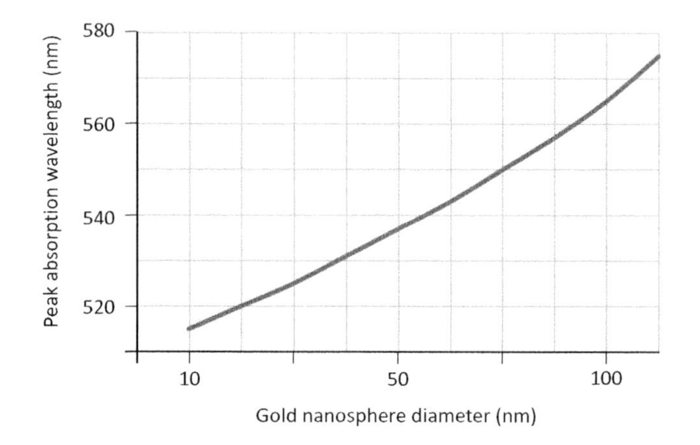

Figure 2.12 The variation in the peak absorption wavelength of the electron plasma as a function of the gold nanosphere diameter. A suspension of 10 nm nanospheres in water, or in glassware, looks 'reddish' whilst 100 nm spheres appear more towards the blue since now the peak absorption is in the red.

Much of the story here is concerned with the ability of structures to produce what are, in effect, wavelength filters. Examples of such phenomena occur extensively in nature – butterfly wings are amongst the most common and striking. This comes initially as something of a surprise: let's make something a different shape and it will appear as something different. However, a few thoughts on electronic circuits shows how we take this for granted. As far as a radio frequency wave is concerned, a piece of straight wire appears as an object very different from a coil of wire yet both have dimensions significantly less than a wavelength. What has become known as 'nanophotonics' has been described as a manifestation of this familiar concept – so here light becomes an electromagnetic wave but in the sense of an electric current.

2.4 Summary

Is light a wave, a particle or a ray or an electric current? The answer is yes to all four possibilities, but it all depends on the context. The preceding discussion has given a glimpse of all these aspects and exemplified some of the circumstances in which light is more readily perceived using one or the other model. There is some interchangeability but usually an appropriate combination of points of view is the simplest way to arrive at a usable understanding. It is quite a fight, say, to model the refractive index through quantum mechanics and particles, but in general bringing one model into the domain of another, as Nils Bohr did with his hydrogen atom model, can give useful, workable and, with care, accurate results. Of course, those of a philosophical nature should consult examples of the many extensive texts on the nature of light.

2.5 Problems

These problems could involve collaborative searching on the internet.

1. What features determine the colour of the objects around you? Consider, for example, butterflies, plants, the bloom on a camera lens or some spectacle lenses, the iris of your eye.
2. What are the light sources around you – for lighting in your room, for your DVD player, in your fluorescent watch, in a 'black lighting' tee shirt, in the sun, in the TV screen and in those old cathode ray tube TV sets? How do these sources vary in structure, in their physiological effect on the observer, in beam quality etc.?

3. What determines the colours you perceive when looking at a particular object? Why do colours apparently differ when viewed at different times of day, or when one is using different light sources or when one goes abruptly from the dark into a brightly lit room? How, for example, could the texture and colour of a wall make a huge difference to the brightness of a room as sunlight comes into it – what might be the mechanisms at work? You should take into consideration the way in which the eye responds to different colours and levels of light.
4. Think about a planar interface between two materials with refractive indices, n_1 and n_2, with $n_1 < n_2$.
   (a) Demonstrate that Snell's law relating the angles of incidence and refraction from material 1 to material 2 can be understood by considering the projection of the components of the optical wavelengths of the incident and refracted rays along the optical interface.
   (b) What is described in part (a) is an example of 'phase matching', also sometimes called 'resonant coupling'. Look around and discuss where this general effect may be taking place, both in everyday events and in the technological world in which we live.
   (c) We mentioned the Brewster angle in the text. At this angle, when light traverses a planar interface from a low-index region into a high-index region, the reflection coefficient for the electric field vibrations parallel to the plane of incidence is zero, giving rise to partially polarised light in the reflected beam. Investigate the generalisation of this observation – look up Fresnel reflection on the web – and note not only the changes in reflection amplitude but also the changes in phase.
5. Would a suspension of gold nanospheres (Figure 2.12) appear the same colour in water, air and high-density glassware? If so, why, and if not, why not?
6. (a) Look up typical values for the free electron density in the ionosphere, in gold and in n-type silicon and use equation 2.3 to arrive at the plasma resonance frequencies.
   (b) From this, comment on when and how these materials become transparent to electromagnetic waves and evaluate the significance of your calculations in terms of possible applications.
7. We have indicated in this chapter that there are practical circumstances when the refractive index can be less than 1. This implies a velocity of light which exceeds the velocity of light in vacuum. However, it is widely accepted that the speed of light in a vacuum is the fastest that light can travel. How can we rationalise these two observations? The key lies in the subtleties of phase and group velocities, hinted at in problem 4 above and discussed a little more in Appendix 4.

3
Light Interacting with Materials

Here we shall explore the principal phenomena which occur when light passes through, and therefore interacts with, a homogenous material.

3.1 Linear Optical Materials

The word 'linear' in this context indicates that light emerging from a material is at exactly the same frequency as the light which entered the material. In other words, all the 'springs' in the mass–spring–damper model (the Clausius–Mossotti equation) mentioned in Chapter 2 are operating within their linear, Hooke's law, region. We also mentioned the concept of the real and imaginary parts of the refractive index, respectively describing the propagation delay and the rate of energy loss by absorption. These are, as with all linear oscillatory systems, intimately connected to each other through the Kramers–Kronig relation[1] for the complex refractive index $n + jk$, relating the real part n and imaginary part k:

$$n'(\omega) = 1 + \frac{2}{\pi} \mathrm{P} \int_0^{\infty} \frac{k(\Omega)}{\Omega^2 - \omega^2} d\Omega \tag{3.1}$$

where P is the Cauchy principal value operator. Entire textbooks have been written with this equation as their sole topic – the principal message is that the real and imaginary parts of the refractive index are intimately related and that by knowing k over the integration range as a function of the optical angular frequency Ω, in principle it is possible to calculate n at the desired frequency ω.

[1] In terms of the *dielectric constant* $\varepsilon = n^2$, the Kramers–Kronig relation is $\mathrm{Re}(\varepsilon(\omega)) = 1 + \frac{2}{\pi} \mathrm{P} \int_0^{\infty} \frac{\Omega Im(\varepsilon)}{\Omega^2 - \omega^2} d\Omega$.

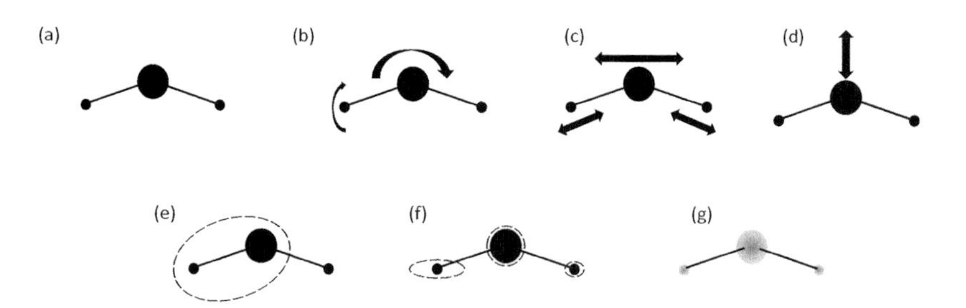

Figure 3.1 Some of the potential resonant modes in a triatomic molecule, indicating the range of rotational, vibrational, orbital and nuclear potential resonances, all of which contribute to the wavelength dependence of light energy, which produces the characteristic colours of materials, their unique spectroscopic signatures. (a) A molecule, e.g. water or carbon dioxide, (b) rotational modes, the entire molecule and bonds rotate, (c) 'longitudinal stretch' vibrational modes, (d) 'lateral stretch' vibrations, (e) vibrations of 'bond orbits', (f) vibrations of atomic orbits, (g) nuclear vibrations in individual nuclei.

There are, as already noted, many resonant modes in a molecular material (Figure 3.1 illustrates some of these). The principal modes of interest here are:

- the various stretching and rotational modes that are possible in a molecule, typically found in the terahertz and microwave regions;
- electronic orbital resonances corresponding to excitation between particular electronic energy levels, typically found in the infrared through to the ultraviolet region.

There are also numerous hybrid oscillating modes where some lower-frequency resonances mix into higher-frequency resonances and produce sidebands, thereby giving a so-called 'fine structure' to the absorption characteristics. The whole story can also extend up to the far ultraviolet and X-ray regions and beyond, where portions of the spectrum embrace resonances within atomic nuclei.

This very brief exploration of the possibilities within the resonant spectrum indicates that these spectra are inherently very complex – a mixed blessing, since on the one hand this complexity implies individual identity for a particular molecular structure, even down to the isotope level, but on the other hand intricate frequency signatures require very careful analysis.

Thus far we have only considered one molecule in isolation. When we bring together groups of molecules, as gases, liquids or solids, then the scope for interactions is considerably enhanced. In this context, recall that if any two molecular structures are sufficiently close to interact with each other, then the corresponding energy levels therein begin to split, as illustrated in Figure 3.2. Going back to the classical oscillator model, similar 'splitting' in resonances can also occur in mechanically coupled systems.

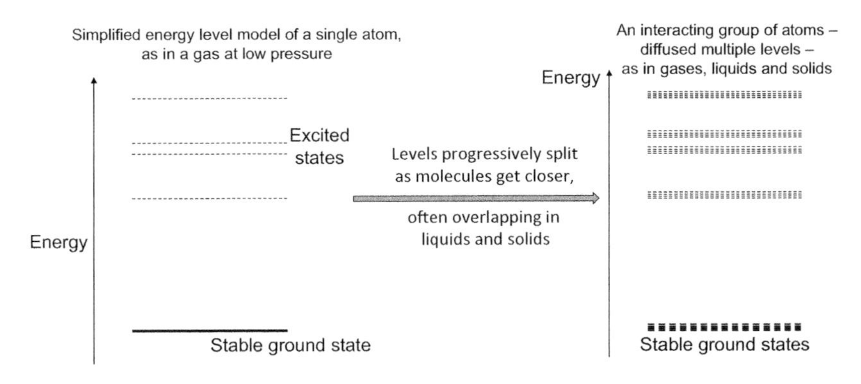

Figure 3.2 A schematic of the splitting of energy levels as atoms get closer to each other. For most solids and liquids these levels begin to overlap into energy bands. Molecules also have similar energy level diagrams – but with significantly more detail therein.

In groups of molecules there are also other forces at work, most noticeably in gases, associated with changes in pressure and temperature. Increases in pressure obviously involve increases in molecular density and consequently also increases in the value of the refractive index. Additionally, bringing molecules closer together increases the level splitting, and thereby broadens any particular atomic or molecular energy level. Bringing molecules together also increases collision probabilities, thereby broadening the perceived linewidth.

Increasing the temperature speeds up the molecular motion, thereby increasing the mean velocity as the square root of the absolute temperature. Hence, because of the Doppler shift, the perceived range of energy levels participating in a particular transition also increases – that is, the absorption line broadens again. However, with gases there is another factor at work. The preceding argument applies to a constant-volume rise in temperature. If the volume of the gas is unconstrained, the answer is somewhat different. The number density of molecules decreases, with the opposite effect.

These pressure- and temperature-related phenomena are present in the spectra of liquid and solid dielectric materials but are far less evident than for gases, except during phase transitions from gas to liquid to solid or, in some cases, when solids or liquids move from one particular molecular state of organisation to another.

The linear propagation of electromagnetic waves in conductors is also important, even in the range of the spectrum where photon interactions become significant. We have previously mentioned the extrapolation of the Claussius–Mossotti equation into the free electron case appropriate to conductors. Here the restoring force is generated by the separation of fixed and mobile charges, giving rise to the concept of collective oscillations at a plasma frequency, above which the metallic conductor becomes electromagnetically transparent. Conceptually, an incoming electromagnetic wave above the plasma resonance can be thought of as reversing the directions of free electrons before they have had time to collide. The collision process is the principal source of conductive losses, so these decrease dramatically above the plasma resonance. Below this frequency, the conductor will typically reflect lightwaves incident upon it, with a wavelength dependence determined by the molecular resonances within the conductor itself. These are observed through the corresponding absorption spectrum but with some losses due to surface-induced currents within the conductor. These induced currents within the conductor penetrate to a depth δ determined by the classical skin effect:

$$\delta = \sqrt{\frac{2\rho}{\omega\mu}} \tag{3.2}$$

where ω is the optical angular frequency, ρ is the resistivity and μ is the permeability of the conductor. Strictly speaking, this approximation is applicable for $\omega \ll 1/(\rho\varepsilon)$, where $\varepsilon = n^2$ is the dielectric constant. At frequencies well in excess of this value, the skin depth δ approaches an asymptotic value, $\rho\sqrt{\varepsilon/\mu}$.

Provided the optical frequency is sufficiently below the nominal plasma resonance the low-frequency values apply (see Appendix 3 for more detail). The skin depth as a function of wavelength for gold and aluminium is shown for reference in Figure 3.3. This is effectively another way of representing an equivalent concept to the imaginary part of the refractive index. If, however, our metallic layer is thinner than the skin depth then some fraction of an incident wave will propagate through it. In conventional terms, the metal conducts. Note, however, that there are significant losses. To give some scale to these losses, the broken and double-dotted line shows the resistance in ohms of a wire one skin depth in diameter and 0.1λ in length.

In the optical region, skin depths are typically in the tens of nanometres; layer thicknesses are easily controlled through straightforward vacuum deposition processes. The currents in a metal layer much thinner than the skin depth are often referred to as surface plasmons. Through similar reasoning, if the dimensions of the metallic sample become small compared with the skin depth then we end up with, in effect, a conductor – a piece of wire. We are now in the domain of photonics referred to as 'plasmonics', one part of 'nanophotonics'.

We have thus encountered the principal features of the optical properties of homogeneous materials. Measuring these optical properties gives us a valuable insight into material structures, and the colour of everything we see is determined through the absorption spectra of the materials around us – what is not

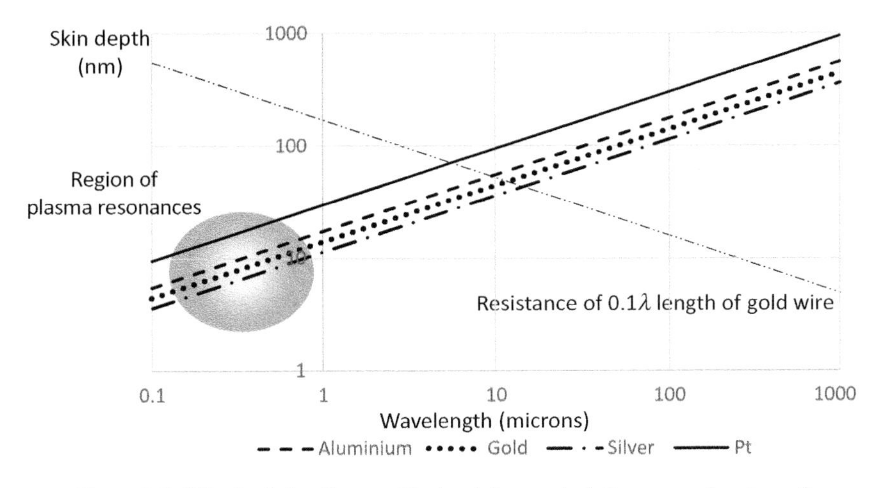

Figure 3.3 Skin depth in silver, gold, aluminium and platinum as a function of wavelength. The last of these has significantly higher resistivity, so the skin depths are substantially higher. The line with negative gradient shown as — · · — · · — · · – gives the resistance in ohms of a gold wire 0.1 wavelength long and one skin depth in diameter.

absorbed is either transmitted or reflected and it is these reflected remnants of the visible spectrum which characterise the world we see. Finally, a word of caution. The optical properties of the things we see are also, as we shall explore in Chapter 4, influenced by their structure: witness the butterfly wing or indeed the sight of oil on water.

3.2 Non-Linear Optical Materials

Non-linear effects occur when the optical frequency emerging from a material contains components which were not present in the optical input. We could interpret this as occurring when our 'spring', in mass–spring–damper model (the Clausius–Mossotti equation), goes beyond its elastic limit. However, it is usually more convenient to interpret non-linear optical phenomena through photons and energy levels.

The optical Kerr effect is the exception. It applies to all materials and is present for all levels of optical power density input. The Kerr effect is the phenomenon in which the refractive index of the material through which an electromagnetic wave is transmitted undergoes a slight change that is dependent on the electric field E_{opt} and is given by

$$\Delta_n = \lambda K E_{opt}^2 = n_2 I \tag{3.3}$$

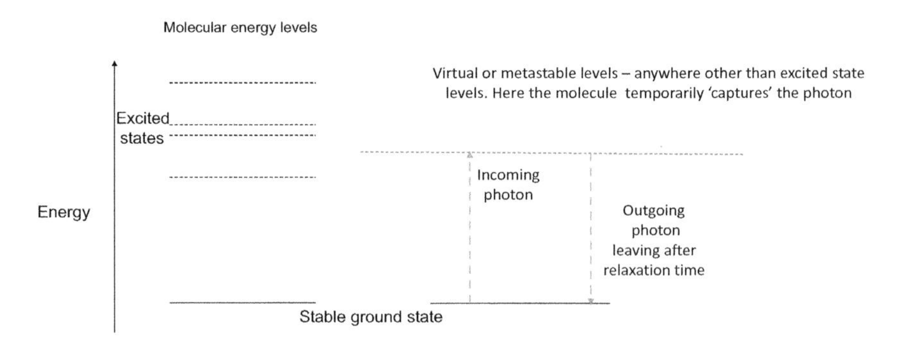

Figure 3.4 The concept of 'virtual' energy levels, where an incoming photon can temporarily excite an atom or molecule to a level not within the energy level spectrum. The photon will escape a short time later. This is a quantum mechanical equivalent to the Clausius–Mossotti oscillator.

where K is known as the Kerr coefficient, n_2 is the Kerr-induced non-linearity, and I is the optical intensity, typically measured in W/cm^2. At optical frequencies this effect can often be ignored but there are sometimes surprises, e.g. in optical communications and precision optical measurement systems, where intensity-dependent phase changes can have a profound impact on system performance. (For silica, used extensively in optical fibres, $n_2 \sim 3\times10^{-16}$ cm^2/W.)

The more interesting non-linear phenomena in photonics are best viewed through the energy level approach, however. We have seen how, in the mass–spring–damper arrangement, energy comes in at the original optical frequency, is stored in the 'spring' for a short time and then released. The delay manifests itself as the refractive index of the material. Viewing this in terms of energy levels and photons, our incoming photon temporarily excites the molecule, putting it into an energy level state which does not correspond to any of the molecular resonances and which is termed a virtual or metastable energy state (Figure 3.4). We have already seen that the stable state of the molecule – the ground state – comprises a wide range of closely spaced energy levels characterised by the molecular structure and the influences on the molecule within its immediate proximity. Consequently, there is the possibility that when the molecule drops down from a virtual energy state to its ground state, it may select a different energy level to which to drop. The existence of these many levels means that photons are emitted at different wavelengths, referred to as Stokes and anti-Stokes, corresponding to wavelengths that are longer or shorter, respectively, than that of the incoming photon. There are two sets of such effects giving rise to the emission of photons with a wavelength slightly

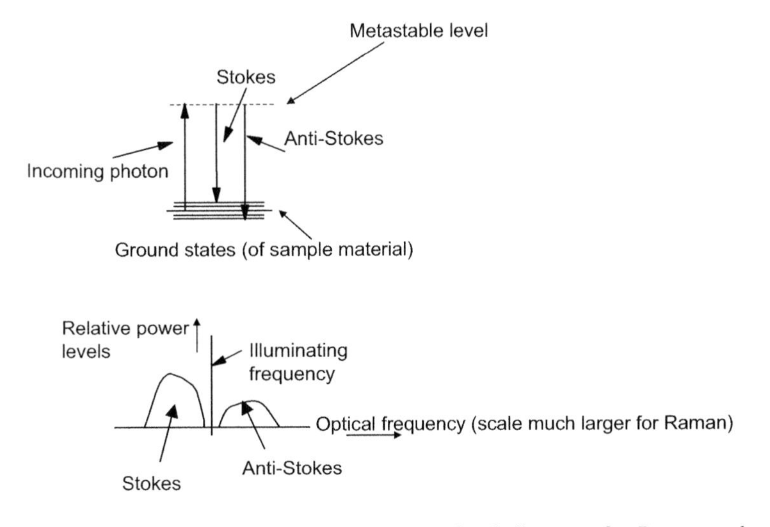

Figure 3.5 (a) The scattering process: energy level diagrams for Raman and Brillouin spectroscopy, where the virtual state releases a photon into a range of different ground states. These ground states are characteristic of the constituent material. (b) A resulting spectrum. The optical frequency scale is much larger for a Raman spectrum, with its greater energy differentials. Because of this, Raman spectra are more easily filtered from the incoming excitation spectrum.

different from the incoming photon, namely Brillouin and Raman scattering. Brillouin scattering is initiated at a lower power density threshold, with corresponding smaller wavelength shifts; for Raman scattering the wavelength shifts are much larger.

This explanation indicates one important feature of these processes: the range of different ground states which can be occupied gives a spectrum of energy differences between the outgoing photons and the incoming photons. This range of differences is uniquely determined by the material through which the light is passing; the effect is present over a broad range of input optical wavelengths and, in particular, at wavelengths well away from the absorption spectrum of the sample. Here, then, we have another optical means for characterising materials, namely examining the Brillouin or Raman spectra and, thanks to the greater photon energy separation between input and output, Raman spectra are more useful (Figure 3.5).

This gives rise to an important question – what happens to the energy differential between the incoming and outgoing photons? The answer is that it released as, or sourced from, phonons, that is, thermal vibrations. In the case of Brillouin scattering these phonons create an acoustic spectrum typically in the tens of GHz region. These phonons – very high frequency

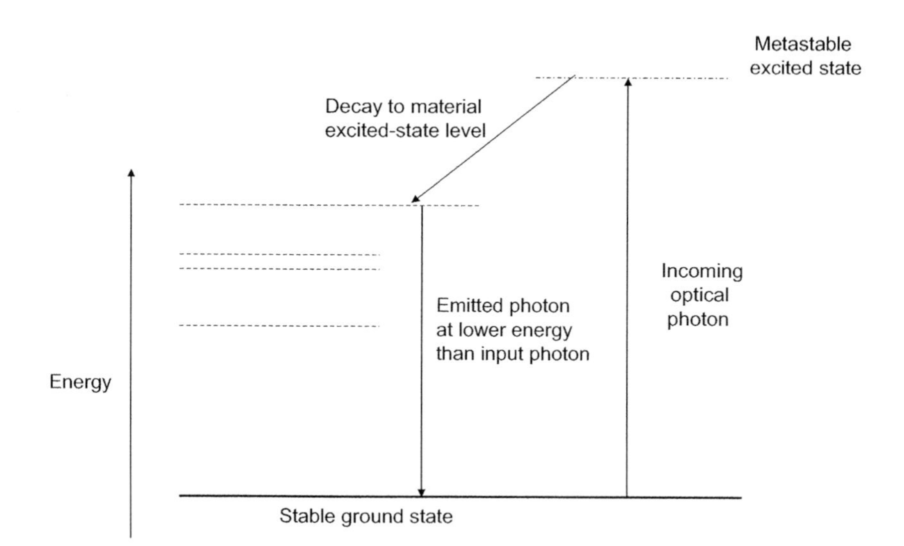

Figure 3.6 Fluorescence spectroscopy. Here the metastable excited state is well defined but has a short lifetime, in contrast with the Raman and Brillouin cases. Decay from it to the material excited state produces heat. Decay from the material excited state produces luminescence.

acoustic waves – are eventually dissipated as heat. For Raman scattering these phonons are in the terahertz region and again finally manifest themselves as heat energy.

Another other important and surprisingly commonplace non-linear optical phenomenon is fluorescence, as illustrated diagrammatically in Figure 3.6. Here a high-energy photon excites an electron to a well-defined transient 'metastable' state and from this state the molecule relaxes into an intermediate level corresponding to an actual excited energy level within the atomic structure. Thereafter, the molecule collapses to its ground state, emitting a photon with a characteristic wavelength. This usually occurs in liquids and solids (only rarely in gases), so there will be a range of energy levels available, giving the fluorescent radiation a characteristic output spectrum. Fluorescence is similar to the Raman effect but had established its identity long before the Raman effect was discovered, in 1928. Also, there is one important difference. In fluorescence the incoming photon excites the molecule into a short-lived but well-defined energy level. This photon relaxes via the two-stage process mentioned above into states related to the complex refractive index of the fluorescent material. For the Raman effect the initial transient state can be at *any* level and is not related to the complex refractive index of the material concerned. Additionally, Raman scattering occurs for all materials, but fluorescence for relatively few.

The overall results are very similar, though fluorescence is the more familiar, from black lighting, where ultraviolet lights trigger visible fluorescence in clothing and décor in nightclubs, to glow worms, which excite their radiating molecules through biological rather than a directly optical interaction, in a process usually referred to as bioluminescence. The basic process is the same – an input stimulus, in the glow-worm context, biochemically excites a molecule to an intermediate state prior to its dropping a little energy before returning to its ground state and emitting a corresponding photon. There are many other variations on the basic theme. Electro-luminescence is perhaps the most familiar, where an electric current excites atoms which thereafter relax and emit photons; this happens in, for example, a light emitting diode. All these variations, however, have one common feature in that the light emitted is characteristic of the material from which it originates and consequently can, among other possibilities, be used as an optical signature in materials analysis.

3.3 The Diversity of Optical Absorption

From straightforward experimental observations, we can conclude that if light – from the sun perhaps – shines on a material, the energy of the light which bounces off that material almost always adds up to a lower total than that of the light which arrived. Consequently, by the conservation of energy, the remainder must be converted into something different, and the question is, what?

The most obvious answer is heat. A black sheet of card left out in the sunshine soon feels warmer than a white one placed beside it. Both sheets will, however, take time to stabilise to their eventual steady temperature – a temperature at which the light absorbed is exactly balanced by the heat lost. This in turn is a function of the thermal conduction and convection routes from the sample and its thermal capacitance. Since the samples are relatively large, the time taken for this equilibrium to become established is measured in seconds or minutes. This in turn leads to the question, what happens if we switch the light off again before this equilibrium is established? The answer is that everything cools. Thermal expansion is happening too, so in this process the actual physical size of our sample increases while exposed to the light and decreases when the light is removed.

There are some interesting implications if we change the context slightly to that of a tightly focussed beam, usually from a laser, incident on a large metallic sample and switch this beam on and off very rapidly (Figure 3.7). We know, thanks to the perceived colour of the metal, that invariably some incident light will be absorbed. We also know that this absorption will take place within one

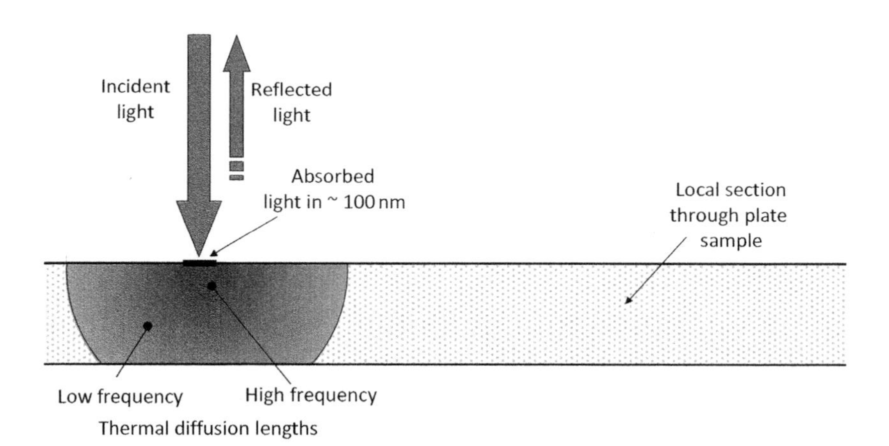

Figure 3.7 Photoacoustic conversion. The light is absorbed in the optical skin depth, producing a thermal wave whose characteristic skin depth is dependent upon the frequency at which the light is modulated.

optical skin depth, which is usually (Figure 3.3) less than 100 nm. All the heat is initially generated in this tiny volume but rapidly diffuses away into the bulk material. The thermal conduction processes are exactly analogous to the electrical conduction process so that, again, if the dynamic heat source (our incident light beam) is switched on and off at a particular angular frequency ω, at this frequency the thermal diffusion depth will be given by

$$\delta_{th} = (2\sigma_{th} / (s_{th}\Delta)^{1/2})/\omega^{1//2} \tag{3.4}$$

where σ_{th} is the thermal conductivity, Δ is the density and S_{th} is the specific heat. Values of the thermal diffusion depth for brass, which is typical of many metallic materials, are shown in Figure 3.8 for a range of frequencies ω up to around 100 MHz. Acoustic wavelengths for compressional waves in a typical metallic material are also plotted for comparison. We can then see that if the light beam is switched on and off sequentially then the local temperature will rise to a specific depth, which, in this range of acoustic frequencies, is much less than the acoustic wavelength. This will induce local thermal expansion, resulting in pressure differentials and hence ultrasonic waves. Such laser-generated ultrasound is an extensive topic in its own right, with important applications in acoustic imaging and non-destructive testing, and numerous full texts describe its principles and applications.

Thus optically induced localised heating, typically from a short-pulse laser beam, produces a highly localised temperature rise, initially within the optical skin depth and rapidly thereafter spreading further into the target material. This leads to another observation, that the smaller the laser spot and the greater

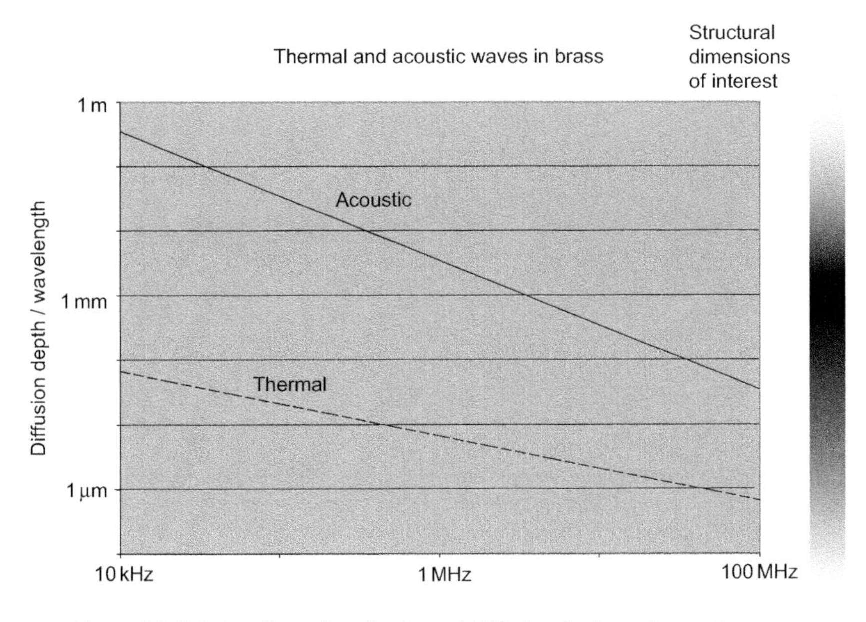

Figure 3.8 Relative dimensions for thermal diffusion depths and acoustic wavelengths, here given for brass.

the energy density applied within a given time period (i.e. the higher the peak power density), the higher the localised temperature rise within the optical skin depth. This highly localised short-term temperature rise can be high enough to melt or even evaporate the metal.

An important application is laser machining, now extensively used in many production processes. For some materials, notably polymers, this machining process may not necessarily involve melting the target but, rather, introducing light at a high enough photon energy to break the chemical bonds which hold the polymeric molecules together.

Closely related to this is laser-induced breakdown, typically but not exclusively in gases, which can give a highly intense source of light stimulated by tightly focusing a laser beam in the sample material (Figure 3.9). In this case breakdown is induced by the large electric field within the light beam at the focus, similar to the arcing sometimes seen around high-voltage power lines. This laser-induced breakdown produces a spectrum which is characteristic of the material in which the breakdown occurs. Laser-induced breakdown spectroscopy (LIBS) is a spectroscopic tool with an immense range of application spanning atmospheric analysis to the characterising of biological cell structures.

Turning light into heat on a less dramatic scale also has its applications. Indeed, thermally driven solar power can provide hot water in warm countries. Large-scale arrays of mirrors focussing the sun onto a black boiler structure, a

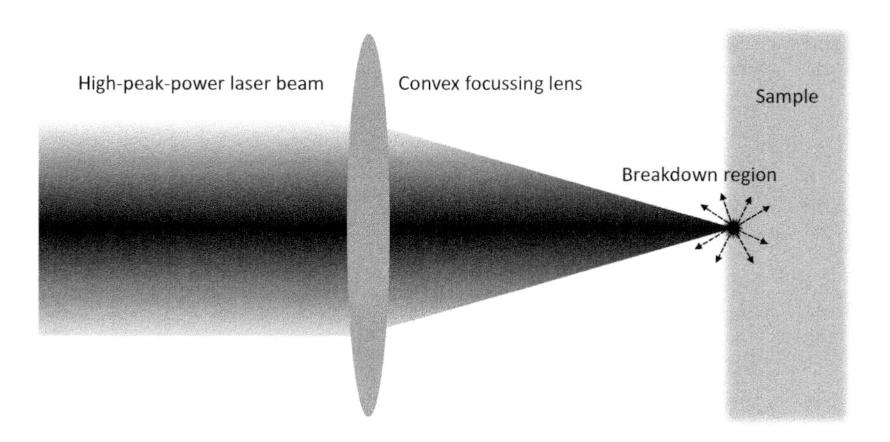

Figure 3.9 Laser induced breakdown spectroscopy, where a tightly focussed laser beam produces very high, very localised electric fields which ionise the sample into a gas, producing a characteristic, usually line, spectrum.

solar furnace, can even provide steam for electrical power generation. In a very different context, imagine that you could fabricate the perfect 'black body', which would absorb all photonic wavelengths. The temperature rise in such a body would then be proportional to the optical power incident upon it. If this thermal change were coupled into an electrical thermistor or a thermocouple then measuring the temperature change would give a direct reading proportional to the optical power; this is realised in practice in a device called a bolometer, which has applications across the electromagnetic spectrum.

In other forms of photodetection, incident light to creates an electric signal, usually a current, proportional to the optical power absorbed. The origins of this direct conversion lie in the photoelectric effect, which led to the concept of the photon via the idea of a characteristic 'work function' necessary to eject an electron from a cathode material into a vacuum.

The photomultiplier (Figure 3.10) was thereafter for many years the principal means through which optical signals could be turned into proportional electric currents. However, it is important to note that in general one photon creates only a single electron, so the resulting current is proportional to the photon arrival rate rather than the optical power. It is also important to note that the idea of the photomultiplier is that each photon-generated electron produces many electrons via the electrode cascade. This process is statistical and introduces noise into the detected current. There have been many variations on the photomultiplier theme. For example, early television cameras were based upon the vidicon, a device which operated through a directed electron beam and a photocathode on which an image was projected.

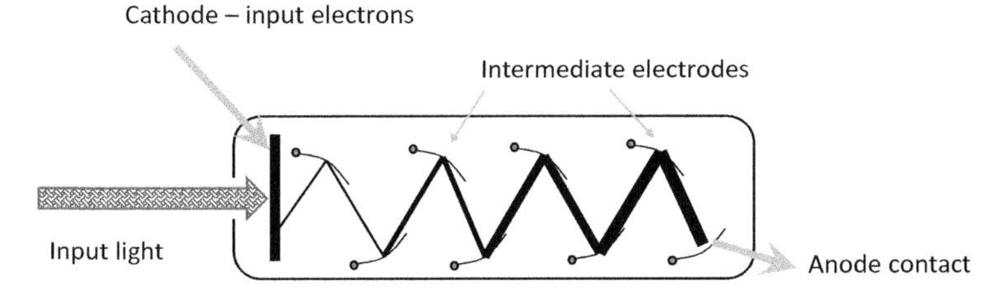

Figure 3.10 The principle of the photomultiplier tube. The input light causes electrons to escape from the cathode and these are multiplied through successive intermediate electrode stages.

In 1947 the semiconductor transistor was invented; this, it is safe to say, changed everything. The semiconductor photodiode now detects optical power in CD and DVD players, optical fibre communication systems, responsive illumination systems, camera systems and a host of other areas.

The basic principle of the semiconductor photodiode requires that the incident photon has an energy exceeding that of the bandgap in the semiconductor (which is comparable with the work function of a photocathode). The photodetector is usually reverse biased (Figure 3.11) and this reverse bias sweeps away the free electrons generated by incident photons in the conduction band and the free holes in the valence band, to produce an electrical current proportional to the incident optical power (remember, though, we are really looking at the incident photon arrival rate, as for the photomultiplier).

This observation leads immediately to one of the principal features of semiconductor photodiodes (and photomultipliers) – there is an optimum wavelength, corresponding to photon energies somewhat above the threshold, where the sensitivity of the photodiode, measured in amperes per watt, reaches an optimum. This can be understood by recalling that each photon can only create one electron–hole pair and each pair will need available energy levels in the valence and conduction bands respectively. If the photon energy is too close to the band edge, the number of available states will be limited so that the absorption process will be relatively inefficient. If the photon energy is too high then, yes, there will still be many states available but each of these higher energy photons will still only produce one electron–hole pair. Consequently, the current produced for a given power level will be reduced, since, for a given total power, the number of photons per second arriving at higher photon energies will be less. This results in a responsitivity plot which depends upon the semiconductor material; it is shown for some common examples in Figure 3.12.

There are other factors which determine the optical detection efficiency (the percentage of incident photons converted to electron–hole pairs) of the

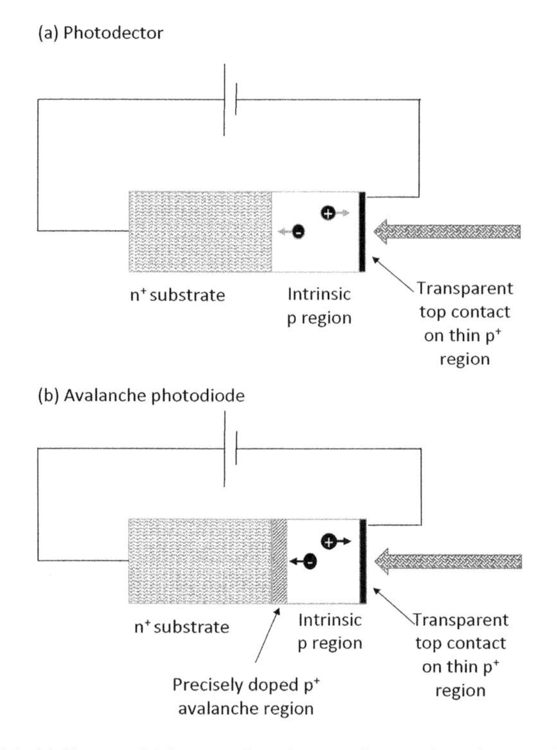

Figure 3.11 (a) Reversed-bias p–n junction used as a photodetector. In (b) we have the avalanche photodiode, which is very similar except that there is a high electric field region near the n^+ region. 'Intrinsic' here means very low p-doping levels.

semiconductor p–n junction photodiode. The most important of these is the depletion length, the length of the region swept clear of carriers in the absence of any photon flux. Any photons absorbed within the p^+ type or n^+ type material on either side of this region will create additional carriers. However, these carriers will find themselves in a very low electric field and so will not separate significantly before recombining, so that these photons have been 'lost'. Nothing comes without compromise and so, in the case of the photodiode, the wider its region, the longer it takes for carriers created within this region to arrive at the ends of the region. This delay affects the available operational bandwidth of the photodiode, which is restricted by the speed with which carriers cross this depletion zone.

The carrier recombination process mentioned above also has some important features. The band structure of a semiconductor is a little more complicated than that hinted at earlier. There is structure not only in particle energy but also

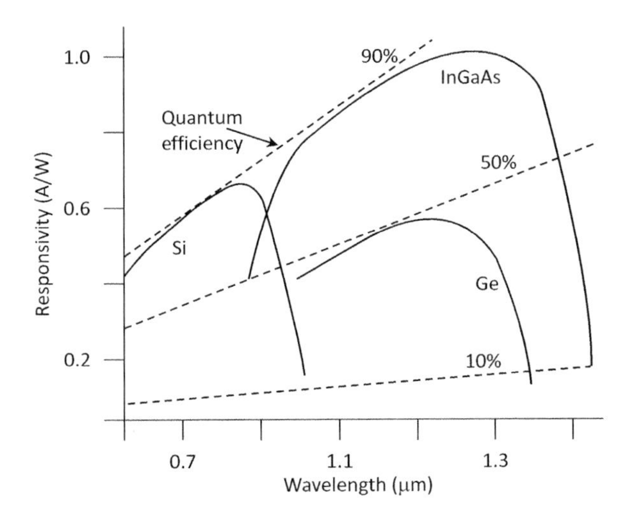

Figure 3.12 Responsivity in amperes per watt as a function of wavelength for three typical photodetector materials: silicon, germanium and indium gallium arsenide. The quantum efficiency is the ratio of free electronic carriers produced to the number of incident photons.

in particle momentum, as indicated in Figure 3.13. In order to generate an electron–hole pair the incident photon needs to match both the energy and momentum states available within the semiconductor band structure. This is particularly evident in silicon – initially the most common photodetector material – where the narrowest bandgap in energy corresponds to a shift in momentum also. The absorption process then has two stages. The photon lifts the electron into 'no man's land', as indicated in Figure 3.13; then the thermal energy within the semiconductor lattice 'kicks the electron sideways' into a vacant spot in the valence band, in order to match the energy and momentum released by the photon. In contrast, gallium arsenide (and the many derivatives thereof; InGaAs, a much used photodetector material, is one) has a 'direct' energy gap, so that the absorbed energy from the photon dislodges an electron from the valence band which goes directly into the conduction band. The recombination processes follow similar paths, with profound effects on the output from this recombination process, which we shall explore later.

There are also two major groups of photodiode detectors that depend on the operation of the intrinsic region. Thus far we have implicitly been discussing the p–i–n photodiode detector, version, with heavily doped p and n regions sandwiched between an intrinsic layer, as illustrated in Figure 3.11(a). The intrinsic region is easily swept free of carriers, and so it is the major optical absorption zone in the photodiode. The electric field in this region can be

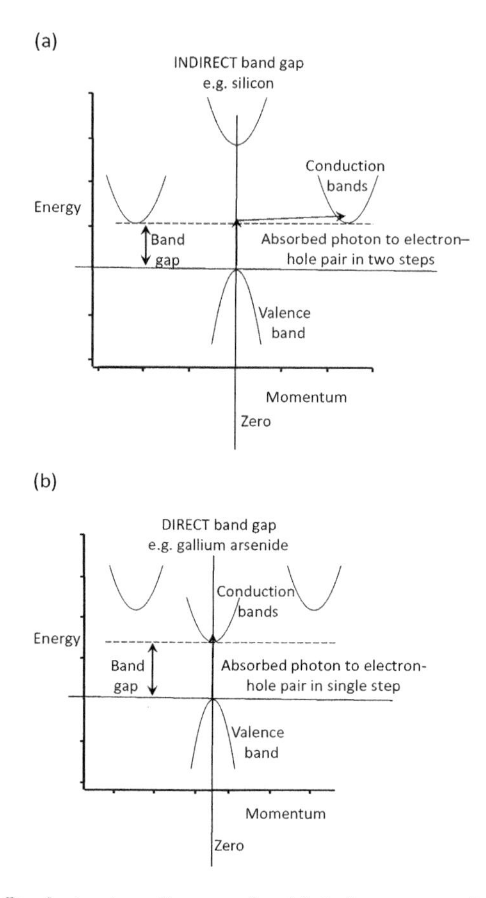

Figure 3.13 Band structure diagrams for (a) indirect-gap and (b) direct-gap materials.

raised to a relatively uniform and high enough level to ensure the fastest possible removal of any free carriers which have been generated. However, there is only one electron–hole pair for each absorbed photon.

Also shown in Figure 3.11 is the principle of the avalanche photodiode. Here we again have a short p^+ region through which, as with the p–i–n, photodiode the input photons arrive, followed by a relatively low-field intrinsic region. However, immediately before the n^+ substrate contact region there is a short, carefully controlled, p^+ type region. An appropriate voltage applied to this structure creates a high electric field zone in the region just before the free carriers (here electrons) arrive at the n^+ (positive) contact. The design of these diodes is quite tricky, since the idea is that in the high-field region the electric field is just sufficient to cause controlled multiplication (but avoiding total

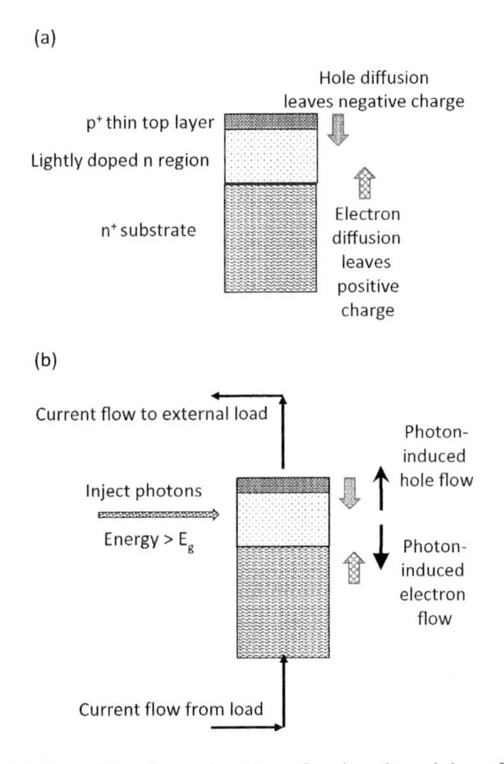

Figure 3.14 (a) An unbiased p–n junction, showing the origins of the built-in voltage. (b) This is changed by providing an external load and injecting photons of energy exceeding the bandgap E_g to produce a flow of carriers for a solar cell. In both diagrams the lightly shaded arrows indicate natural diffusion into the depletion layer.

breakdown) of the input electrons arriving from the intrinsic region, therefore increasing the detected current by a controlled factor. There is inevitably some statistical variation within this process, as with a photomultiplier, so the avalanche diode is subject to additional current fluctuations. These are apparent as additional noise. However, the extra sensitivity garnered from the use of this process often far outweighs the noise penalty. The light sources themselves also carry their own inherent and inevitable noise component, as will be discussed in more detail in Chapter 5.

Thus far, we have briefly explored the action of the p–n junction as a photodetector when operating under reverse bias. However (Figure 3.14), even when unbiased from external sources, a p–n junction sets up its own depletion layer and its own in-built, albeit small, reverse bias: the free electrons in the n-type material diffuse into the p-type material and the holes move vice versa.

This zero-bias voltage is typically in the region of a very significant fraction of one volt. Suppose that now we have light shining on the depletion region. Again there will be some electron–hole pairs generated within the depletion region and a new equilibrium state will be established. This equilibrium state becomes particularly interesting and, indeed, highly relevant if, as indicated in the figure, we connect a resistive load around the outside of the illuminated semiconductor device. The new equilibrium state will then establish itself as an electric current proportional to the photon flux absorbed within the junction. The p–n junction has now become a solar cell, an increasingly important device for the generation of clean electrical energy.

This absorption of photons to create free electron and/or hole carriers in semiconductor materials, has another manifestation even when no external bias is applied and there is no p–n junction. The electron–hole pairs produced by photon absorption generate an electric field in the semiconductor material and that field will persist until the carriers recombine. Thereafter what was originally photon energy will be released as heat. However, whilst the electric field is there, both the field itself and the presence of the additional carriers in the material introduce tangible mechanical forces (nothing whatsoever to do with the thermal forces discussed earlier). This leads to a phenomenon known as photostriction, in which a mechanical strain is introduced that is dependent upon the incident photon flux and also on the absorption depth within the material sample, resulting in a greater force on the illuminated side. In the case of silicon this photostrictive strain is negative; the material shrinks very slightly when illuminated. In germanium the effect goes in the reverse direction. The photostrictive effect has also been observed in many other materials not commonly thought of as photoelectric. Of these the most important class is the perovskites, which are more commonly associated with piezoelectricity, the generation of electricity by pressure on a crystal (and also, more recently, solar cells). However, light does introduce free charge carriers in these structures, and these free charge carriers modify electric fields and thereby chemical bond strengths in ways which are varied and quite complex. In all cases the photostrictive effect is relatively small, so sensitive mechanical structures need to be devised to observe its impact.

Thus, the absorption of light in materials has diverse and practically important results. The process can generate heat, with implications ranging from solar furnaces to laser generated ultrasound. It can also generate free electrical carriers – electrons and holes – resulting in electric currents that can be used both for measuring light and for generating electricity. The free carriers can also create mechanical changes simply by their presence, through electrostatic forces interacting with the overall material structure and also with the chemical bonds within the structure. Furthermore, the wavelength dependence of absorption in solids has enormous application in the spectroscopic analysis of materials and material structures. Additionally, there are some very interesting consequences when the absorption process is viewed in reverse, namely optical emission phenomena, which we shall briefly cover in the section which follows.

3.4 Light Emission

Light emission from sources other than the sun, invariably associated with thermal emission, has been an important contributor to human evolution for half a million years or thereabouts. From a spectroscopic perspective such light has properties determined by the 'blackness' of the burning body, its temperature and the spectroscopic photoemission properties of the material which is burning. For example, the classic 'table salt in the gas flame' demonstration graphically illustrates the yellow sodium lines.

A perfect 'black body' is an idealisation of a structure which perfectly absorbs all electromagnetic radiation at all frequencies. Since it reflects nothing, it is black. When heated to an absolute temperature T kelvins, this black body produces a spectral density measured as the power per unit area radiated per unit solid angle, per unit wavelength, $B_\lambda(\lambda, T)$. This is given by Planck's law:

$$B_\lambda(\lambda, T) = \frac{2hc^2}{\lambda^5[\exp(hc/\lambda k_B T) - 1]} \quad \text{in W}/\left(\text{sr m}^3\right) \tag{3.5}$$

with spectral characteristics as shown in Figure 3.15. The visible spectrum is also indicated. The position of the peak in the black body spectrum is determined by the Wien displacement relationship,

$$\lambda_{max}(T) \simeq \frac{2900}{T} \tag{3.6}$$

for λ in microns and T in kelvins.

This peak in the radiation spectrum has given rise to the concept of the 'colour temperature' of a source of light. The sun, for example, has an effective colour temperature of 5778 K, corresponding to a peak radiation at about 500 nm – in the green. If in contrast, we look at the earth's surface, at typically 290 K, then the peak radiation at this temperature occurs at about 10 μm. By coincidence, carbon dioxide gas has a strong absorption line at around this wavelength. Consequently, radiation emitted from the 'black body' earth is absorbed by the carbon dioxide gas and this absorption manifests itself as heat. The more

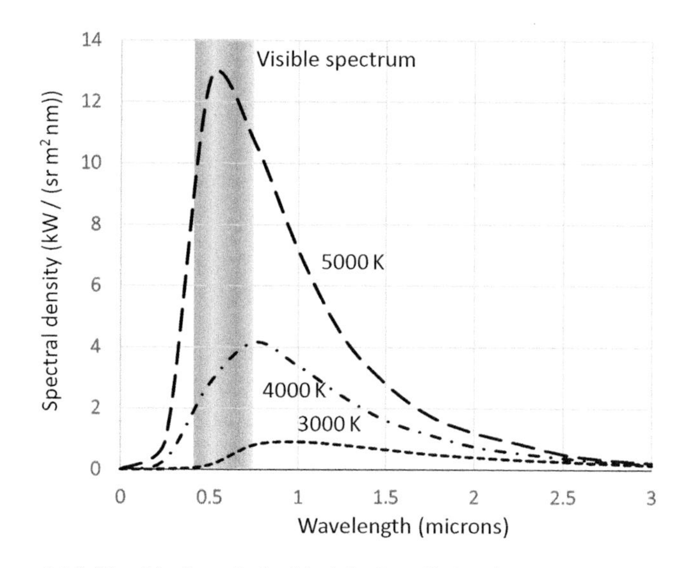

Figure 3.15 Planck's formula for black body radiation from sources at various temperatures with the visible spectrum superimposed.

carbon dioxide, the more the absorption; hence global warming. Furthermore, carbon dioxide is not the only gas which absorbs within this spectral range – atmospheric methane is another contributor, with a far higher (by a factor >20) absorption cross section, but with a somewhat shorter stable lifetime and lower concentration than that of the CO_2 in the atmosphere.

As electricity became more available and understood, the concept of passing electric current through a filament, thereby heating it to produce light, gradually emerged, with early demonstrations from Alessandro Volta and Humphrey Davy consolidating into Edison's carbon filament lamp, enclosed in a weak vacuum to prevent burning of the filament. This well-known invention emerged into everyday use for over a century. It was the best readily available light source despite its inevitably inefficient (Figure 3.15) conversion of heat into visible light: most of the radiated energy appeared in the infrared, heating the immediate surroundings. The colour temperature of an incandescent electric light bulb is typically around 2000 K, which is less than half that of natural daylight, and corresponds to a peak emission wavelength of about 1.5 microns.

Technically, in spite of their huge contribution to human well being over many years, the vast majority of incandescent (i.e. hot-filament) light bulbs do not directly exploit the spectroscopic properties of the materials involved in their fabrication. However, the fluorescent light bulb is deeply rooted in spectroscopic understanding. Figure 3.16 illustrates the principle of operation

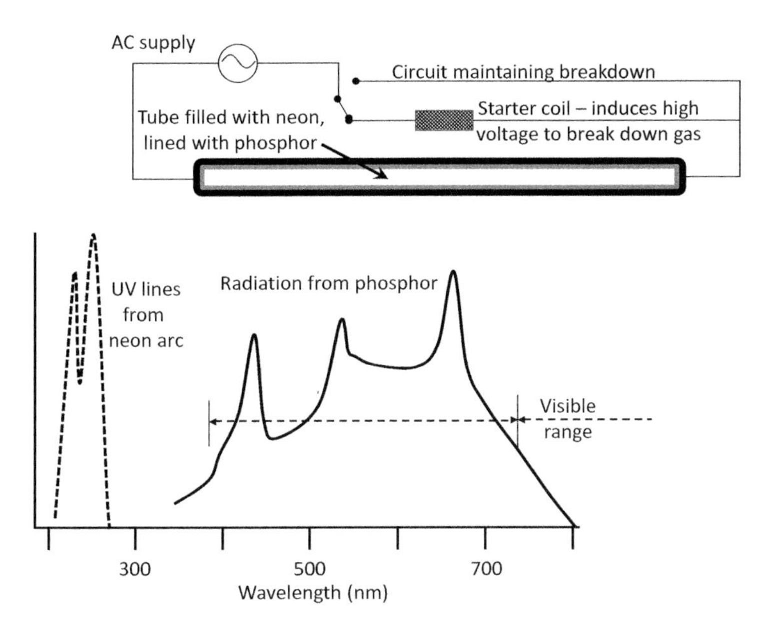

Figure 3.16 The basics of the fluorescent lamp, where a UV discharge introduced by electrical breakdown of the gas (neon) in the tube excites the phosphor to produce visible radiation.

of a fluorescent lamp, indicating the origins of the relatively dull warm-up time before the lamp becomes fully operational. The effective colour temperature, which is determined by the properties of the phosphor, is typically around 4000 K, but the spectrum also includes some sharply defined lines. These lines depend on the combined properties of the gas discharge lamp used to excite the phosphor, the phosphor itself and the local temperature distributions. A typical spectrum is also shown in Figure 3.16. This combination of a source spectrum and a phosphor spectrum, originally conceived over a century ago, continues to find application in today's light emitting diode illuminators.

The light emitting diode, based upon a semiconductor p–n junction, has evolved over the past half century or so into the preferred source of artificial light, thanks primarily to the greater electrical-power-to-optical-power conversion efficiency and the significantly enhanced overall reliability. The principle is relatively straightforward (Figure 3.17). First and foremost, the semiconductor must be a direct bandgap system, to minimise 'non-productive' electron–hole recombination resulting in phonons and hence heat. The direct-bandgap system significantly improves the probability of electron–hole recombination resulting in photon emission. The second principal consideration is that the recombination region must be situated as close as possible to the surface

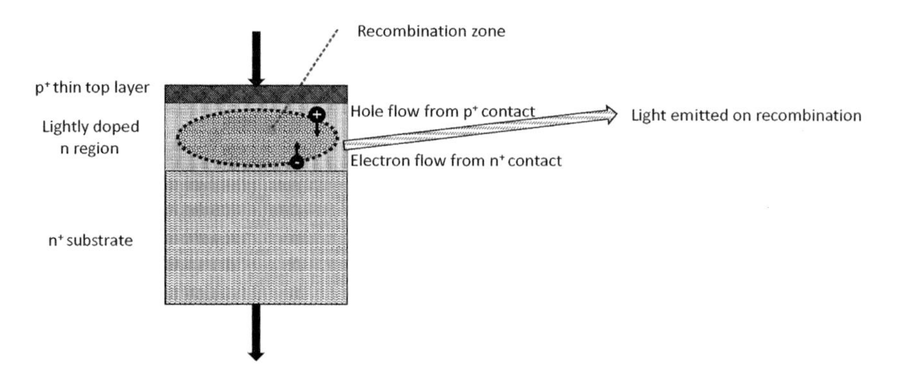

Figure 3.17 The basics of a light emitting diode, in which free carriers combine in the lightly doped n region to produce light.

because, by definition, photons generated through recombination could also be readily absorbed into another electron–hole pair. The structure shown in Figure 3.17 comprises a highly doped n^+ type substrate material with a shallower, much less doped, n-type layer and a thin p^+ top electrode.

When this structure is forward biased (that is, the n^+ type region is negative with respect to the p^+ type region), the electrons emerging from the n^+ region enter the lightly doped n region and there recombine with the holes injected from the p^+ contact. The recombination process produces photons at a wavelength determined by the bandgap of the semiconductor concerned and over a wavelength range determined by the available energy level distributions in the conduction and valence bands of the semiconductor. In most semiconductors this wavelength range is about 5% of the bandgap. Early light emitting diodes were based upon gallium arsenide and consequently, since gallium arsenide has a bandgap of about 1.4 electron volts, were at the extreme end of the red part of the spectrum. Gradual progress in the synthesis of this class of semiconductors, designed to increase the bandgap, eventually produced a blue LED in 1995, recognised by the award of the Nobel prize in 2014.

In principle, the current 20% global electrical consumption in incandescent light bulbs could be reduced by a factor of five through transfer to this technology. Like the fluorescent bulb, the LED lamp uses short-wave-length light (blue and the near ultraviolet) together with a phosphor to produce an apparent white light, similarly to the principle of the fluorescent tube. In common with the fluorescent bulb, much of the visual acceptance of white LEDs depends upon the detailed design of the phosphor and the way in which it is mounted. The phosphor will allow through some of the original blue

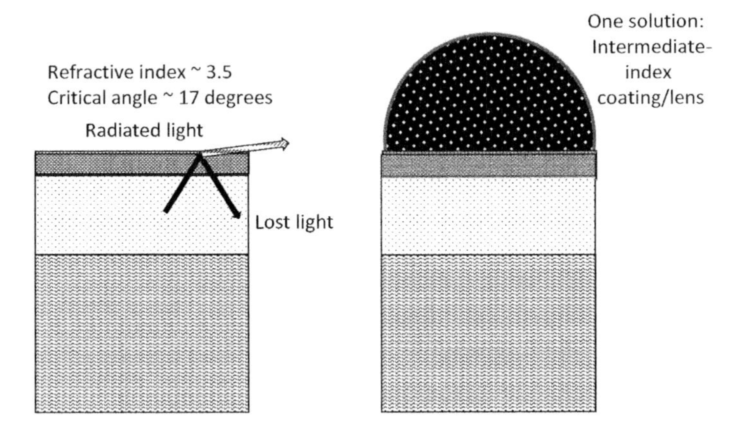

Figure 3.18 (left) Light lost through total internal reflection and (right) a typical solution comprising a surface mounted lens.

whilst generating the longer-wavelength components of white light to produce acceptable illumination.

The light which is generated within the semiconductor material in an LED can radiate randomly in all directions. This indicates one of the basic properties of simple LED structures, namely that, thanks to the principle of total internal reflection, any light incident on the interface between the semiconductor and air (Figure 3.18) at an angle exceeding about 17° (the refractive index of LED materials is typically 3.5 to 4) is reflected back into the structure, where it will eventually be lost due to absorption processes within the semiconductor material. This in turn has led to a variety of ingenious mechanisms designed to circumvent this significant loss; these mechanisms are based upon combinations of lenses, coatings and gratings and result in reasonable efficiencies for LEDs, now typically ~ 100 lumens per watt, and continuing to improve.

This measure of efficiency is, in itself, somewhat strange. The lumen is a physiological unit based on perceived visual intensity. The watt is the input electrical power. Strictly then, any such efficiency values should take into account this somewhat subjective physiological average, which incorporates the visual sensitivity of the eye. This in turn (Figure 3.19) is very different both between individuals and in bright daylight hours (cone vision) compared with the gathering darkness (rod vision). Daylight vision peaks in the green, and there is now an internationally agreed equivalence between lumens and watts, defined for a specific wavelength of 555 nm, for which 683 lumens is taken as equivalent to 1 watt of light at the same wavelength. A value of 100 lumens per

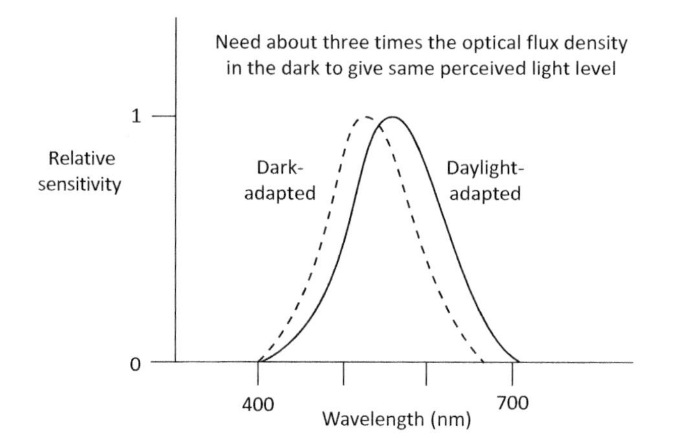

Figure 3.19 The visual sensitivity of the eye showing the shift to the blue for dark-adapted rod vision. Also, the rods and cones have different spatial distributions on the retina, giving different resolutions in bright sunshine compared with those in the gathering dusk.

watt then corresponds to a conversion efficiency of about 15% , far higher than the conversion efficiency of an incandescent bulb.

There is much more to understanding human vision than these comments can provide. However, the different roles of rods for low light vision and cones for higher light levels, and their different spectral sensitivities, provide some clues to the roles of real and artificial illumination in human lifestyles. In particular, the relatively low-level 'warm glow' of incandescent lamps is arguably conducive to a much more restful frame of mind than the much brighter blue-emphasised light from LED illumination. Indeed, as LED 'white lights' becomes more widespread, there is an increasing awareness that such illumination not only tends to disturb sleep patterns but also produces visual confusion when one is moving between dark areas and bright areas illuminated by LED streetlamps. There is much to learn on optimising the balance between spectral distribution, brightness and human physiological and psychological reactions in bringing this application into broader acceptance.

The LED has, however, found wide acceptance after a modest start in calculator displays of just a few digits. Not only is it evolving into a preferred form of illumination, it has also found its place in display technology, from the huge, on the football field, to the tiny, in hand-held devices. The recent emergence of the OLED (organic LED) has added flexible screen technology to the list of possibilities.

There is an important variation on the LED. Thus far we have looked at surface emitters and acknowledged the impact of total internal reflection. If we can confine the radiation within a waveguiding region and arrange for it to come from just one face then, in principle, a much more efficient device is realised. Additionally, if we make the system sufficiently small in cross section, it becomes compatible with optical fibres; the edge-emitting LED is an important light source in fibre communications.

Suppose that we make a relatively minor modification to the basic LED of Figure 3.18 and coat both ends with a wavelength-selective mirror with a very high reflection coefficient (say 99%). This will send nearly all light which is endeavouring to escape from that end surface back into the cavity where the light is generated. We have to ensure that the mirrors are carefully aligned with each other – preferably they should be perfectly parallel. In this case any light which bounces back into the cavity will not escape (Figure 3.20). In doing so some light will be reabsorbed in the semiconductor between the reflective surfaces to create yet more electron–hole pairs. Whilst some elections and holes in such pairs may go their separate ways and be reabsorbed in the contacts, the majority will recombine to form another photon. This creates yet more light, in addition to the light created through the ongoing current injection. But this mechanism only works effectively if the wavelength of this new photon corresponds to the reflection wavelength of the mirrors. Within a very short time the light bouncing to and fro will select the specific resonant frequency dictated by the mirror spacings, and the outcome is the semiconductor laser. The semiconductor laser takes two basic forms, the edge emitter and the surface emitter, and is now essential in optical fibre communication systems, in CD and DVD players and in the ubiquitous laser pointer.

Laser light has special properties, namely spatial and temporal coherence, which express the correlation of the optical phase factor across the beam emitted by the laser and in time. Temporal coherence is often expressed as a *coherence length*, along the direction of propagation of the beam. The *coherence time* refers to the phase correlation along the beam. The coherence time is very short for an LED and much longer for the laser. It is equivalent to the frequency linewidth of the laser oscillator, which quantifies the output frequency jitter over a long observation period compared with the coherence time. It is a reasonable assumption that the coherence time is the inverse of the linewidth in frequency units, though there will be slight variations on this depending on the detailed definitions applied.

Spatial coherence is very low in a light emitting diode, in which individual photons are generated from randomly selected energy level differentials in a emitter and emerge in random directions. However, in a laser the light emitted from the cavity has been selected in both frequency and direction by the

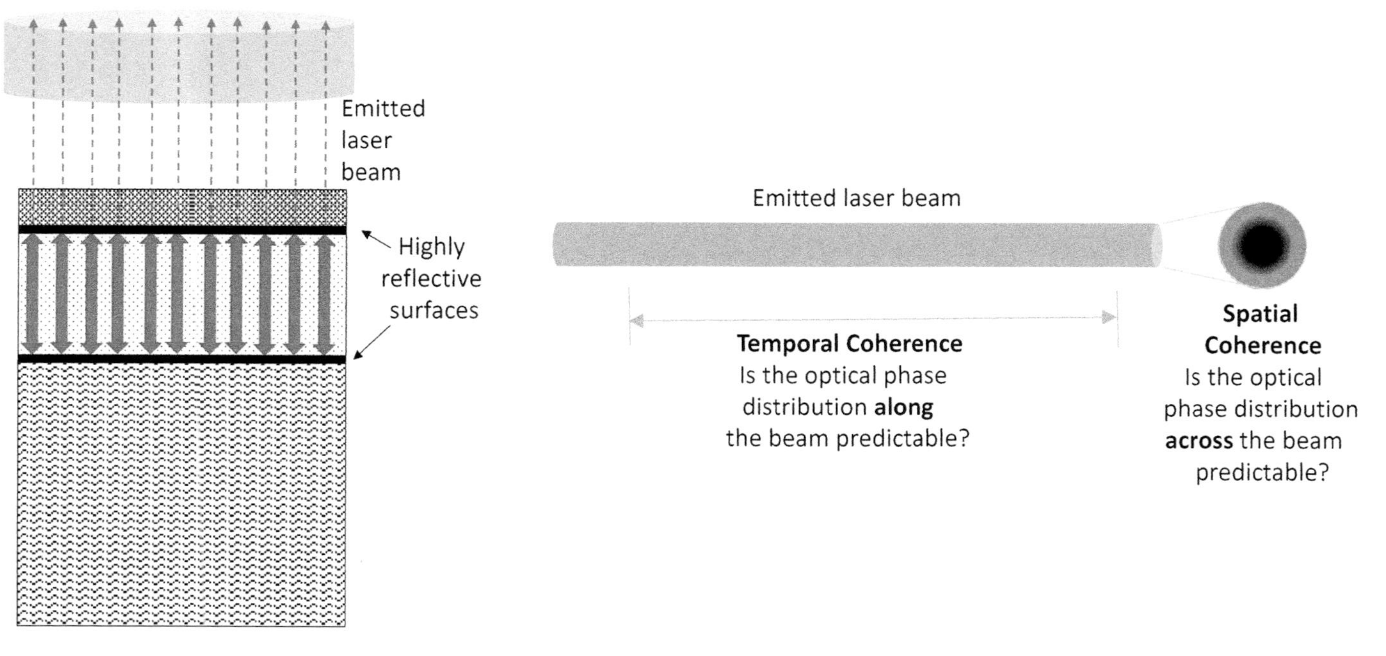

Figure 3.20 (left) Turning an LED into a laser using reflective surfaces and (middle and right) the concepts of temporal and spatial coherence.

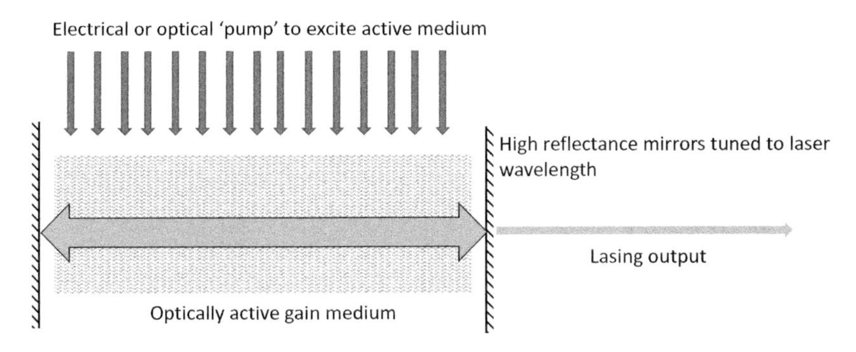

Figure 3.21 The generic schematic of a laser system using external excitation into an optically active gain medium, with feedback from high-reflectivity wavelength-selective mirrors.

characteristics of the resonator itself and the optical-gain medium therein. Consequently, the light emerging is very well correlated in phase across the emitting face; hence the concept of a *coherence area*. In most laser systems the coherence area is identical to that of the emerging beam. Figure 3.20 illustrates the concepts of spatial and temporal coherence.

The same mechanism – surrounding the optically active medium by a suitable resonant cavity – can be applied to any light source which depends upon energy level transitions rather than on thermal radiation, to create a lasing system (Figure 3.21). Electric arcs can excite radiation from gases, and a very bright white light can be absorbed by some materials to produce a specific fluorescent output. The act of enclosing such systems in a suitable mirror structure (selected to the fluorescent wavelength) can give the conditions for lasing. This is exactly analogous to providing a tuned capacitance–inductance electronic circuit around an amplifier in order to inject some of the output into the input and produce an electronic oscillator. We shall encounter lasers many times during the remainder of this book.

3.5 Materials with Controllable Optical Properties

The optical properties of materials depend on two basic sets of parameters. The first set comprises the properties of the individual molecules within that material, which can be understood either through the Clausius–Mossotti relationship or through quantum mechanics and energy levels. The second set of parameters comprises the density of the molecules and their orientations with respect to each other. All these optical properties are essentially encapsulated

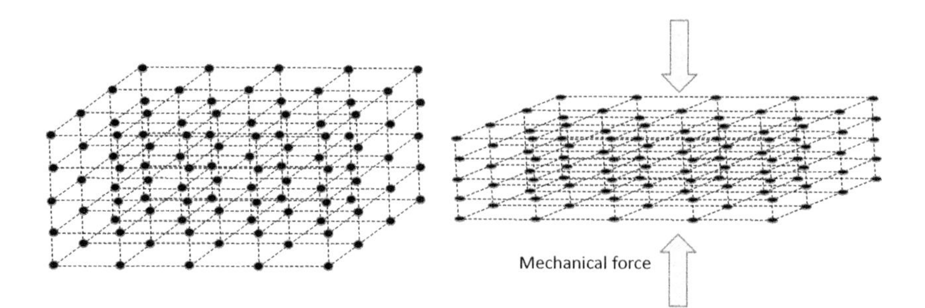

Figure 3.22 How a mechanical force can modify the refractive index of a solid – here shown with a regular molecular structure – by forcing atoms closer together or further apart. Generally, any anisotropic force distribution will, as shown, produce birefringence. External electric fields can produce similar effects in some types of crystalline solid.

in the refractive index, and as we have seen, the greater the molecular density, the greater the refractive index at a particular wavelength. There is, then, one obvious means for changing the index of a material and that is to simply compress it. We have already seen this stress effect in the behaviour of gases under pressure. In solids it is sometimes called photoelasticity and characterised through a stress-optic coefficient.

The stress effect in solids depends upon the way in which the forces are applied. For example, an isotropic pressure brings the molecular structure closer together in all directions whilst a force applied from a single direction (Figure 3.22) compresses the structure in the direction of the force but, as expressed in Poisson's ratio, causes expansion in the other directions. Consequently, light polarised with its electric vector in the direction of the compressing force will see an increase in the index, whilst light polarised orthogonally will see a decrease. Thus birefringence is induced.

This effect could, for example, form the basis of an acousto-optic phase modulator. Here, an optical beam passing perpendicularly to the direction of travel of an acoustic wave in a transparent liquid or solid, and with dimensions in all directions significantly shorter than the acoustic wavelength, will see the index oscillate as the acoustic wave passes through the optical beam. Consequently, the optical beam will experience a variable delay at the acoustic frequency.

By the same token, if we pass an acoustic wave through a relatively large transparent material extending over many acoustic wavelengths we have a travelling periodic variation in refractive index. If we then launch a wide optical beam (compared to the acoustic wavelength) across this structure then, via the acoustic wave, we have the basis of an electronically tunable diffraction grating system, of which more in the next chapter.

Applying an electric field to a molecule or group of molecules can also have an impact on the refractive index of that group of molecules, but this does require a 'cooperative' crystalline structure and cooperative constituent molecules. However, the electro-optic effect, as this is known, is a much more versatile phenomenon than the stress-optic effect. In electro-optic materials either or both of the real (delay) and imaginary (absorption) parts of the refractive index of the material can be modified by the application of a varying electrical signal.

One electro-optic effect which applies to all transparent materials is the Kerr effect, which has already been mentioned. There are, though, many other circumstances in which an applied electric field at low frequencies can affect the optical propagation constants. Of these the best known is the Pockels effect, according to which the refractive index of a material varies linearly in response to an electric field. The Pockels effect does, however, require very specific crystalline structures within the material of interest and these crystalline structures are, of necessity, birefringent; that is, they have different indices for in-plane and out-of-plane polarisation states. Related to this is the possibility of electro-gyration, which applies to crystalline circularly birefringent structures (see Appendix 1). Here the circular birefringence can be tuned in a way that is linearly dependent upon an externally applied electric field. Some sugars, in solution or when purified as solids, are circularly birefringent; dextrose is a well-known example.

These phenomena can all be viewed as electrically induced modifications to the coefficients in the Clausius–Mossotti relationships, which in turn result in changes in optical properties. These phenomena all depend upon the optical wavelength and in particular are highly sensitive around material resonances.

These resonances often coincide with peaks in optical absorption, and consequently there are many routes through which the absorption, of a material around these resonances can be modified. For instance, changes in absorption in semiconductors are feasible and, for some materials, it is possible to electrically modify the perceived colour.

The electro-optic effect is versatile but immensely complex. In contrast with stress-optical effect, which is evident in all materials, the electro-optic effect is only ubiquitous in the case of a relatively low-level Kerr effect. The many other possibilities for changing both the real and imaginary parts of the index under the influence of an electric field do require 'cooperative' material structures and often imply operation in the vicinity of a material resonance and/or within a particular structural crystalline configuration. With careful material selection and structural implementation, this effect can be utilised to make optical phase or frequency modulators, optical intensity modulators, spatial light modulators (as in a digital projector), tunable filters and a host of other useful devices. There are indeed many possibilities!

3.6 Summary

The most obvious optical property of any material is its refractive index. The real part indicates the speed with which an optical phase front propagates (at the phase velocity) through the material and the imaginary part indicates the inherent absorption losses within the material. The real and imaginary parts are, in common with any other linear oscillatory structure, linked through the Kramers–Kronig relationships. These optical phenomena can also be viewed in terms of quantum mechanical energy levels and photon interactions. Both approaches give useful insights, and the choice of approach is, in the final reckoning, a matter of suiting the concept to a particular situation.

These material properties all vary significantly with optical wavelength, as they also do at wavelengths well outside the optical region. This variation provides a unique signature and is an indispensable tool in the characterisation of materials through spectroscopy. There are also non-linear effects, notably, in Raman and Brillouin spectroscopy, which provide useful signatures that are independent of the probing wavelength. However, these do require much higher optical power densities.

The absorption of light in materials also has numerous manifestations: absorbed light becoming heat via a phonon excitation process; re-emergence of the light at a lower photon energy in fluorescence and phosphorescence; electron–hole pair generation and photo-detection; laser driven photo-acoustic effects. These processes all operate, to some extent, in reverse and the optical properties of materials can themselves be modified in response to numerous external stimuli, We have mentioned mechanical forces and electric fields but anything which can modify the local molecular structure and/or the binding forces between nucleus and electrons within that structure will also change the apparent refractive index; consequently, intense heat and nuclear radiation are two other possible candidates for changing optical material properties.

Much of photonics is concerned with understanding the optical properties of materials and applying this understanding to a particular situation. Here we have only mentioned briefly the underlying theoretical analysis necessary for detailed design. However, a basic conceptual understanding is essential in identifying and utilising appropriate optical materials.

In practice, this understanding of materials needs also to be complemented by an understanding of the interaction of light with particular structures which is the topic of the next chapter.

3.7 Problems

1. (a) The sunlight that reaches the outer atmosphere of the earth has a spectrum very similar to that of a black body at about 5775 K (see Figure 3.15). Discuss how the spectrum might appear when it reaches the earth's surface and explain any differences, both in spectral structure and in power density per unit wavelength.
   (b) Bright sunshine reaching the earth has a power density of ~1 kW per square metre. Consider the sunny-day 'magnifying glass onto paper' experiment. What would be a reasonable estimate for the power density in the focussed spot?
2. (a) The value of the Kerr coefficient for silica can be obtained from data given just after equation 3.3. Using this value, estimate the Kerr-induced phase change in a total optical path of 50 km of optical fibre carrying on average 1 mW of power in a circular cross sectional area 8 microns in diameter. The input power is attenuated in transmission – hence we use the average power transmitted. If we were to model this more accurately and take the variation around the average into account, would the total perceived phase delay increase, decrease or stay the same – and why? (You may need to look up an appropriate value for the fibre attenuation.) Additionally, the power density will not be uniform across the fibre cross sectional area; the typical fibre mode shape is roughly Gaussian across the diameter. Using the same logic, will this increase, decrease or not significantly change the impact of the Kerr effect from the original 'uniform in all directions' estimate?
   (b) Suppose two optical signals at different wavelengths are introduced into a fibre at power levels high enough to give significant Kerr effect phase changes. What changes might you expect to find in the spectrum of the output signals, and why? State your criteria for a significant change.
3. (a) There are two basic approaches to detecting light – as photons and when absorbed as heat. Light can also be regarded as an electromagnetic field. For radio waves we use this field to excite movement in electrons in a wire – why don't we do this with light? Or could we also do it with light?

(b) Compare the basic properties of (i) the spectral response of thermal detectors (bolometers) and bandgap detectors and (ii) discuss the factors influencing the intrinsic sensitivity of the two detection processes, endeavouring to derive some comparative performance estimates (the reason why photodetection is dominated by photon detection processes might be a starting point).

(c) Suppose you have a bandgap-based detector material with a bandgap of 1 eV and a quantum efficiency equal to zero above the threshold wavelength and unity immediately below the threshold wavelength. Calculate the responsivity in amperes per watt for such a detector over the wavelength range 0.1 to 2 microns. Explain why this curve looks nothing like the practically observed values in Figure 3.12 and why there is a peak in the responsivity in the curve which you have derived.

(d) You are given the task of optimising the sensitivity of a detector system to visible light (say to 300–800 nm wavelengths, which goes a little beyond the visible at both ends) that is focussed down to a spot. You have no limitations on technological design for the detector. How would you approach this?

4. (a) Estimate the percentage of generated light that actually escapes from a simple planar-surface light emitting diode which produces light through the full 4π steradians solid angle.

(b) There are lensed versions of light emitting diodes mentioned in the chapter – design what you feel would be an optimum lens using ray optics. (Assume that glasses of refractive index up to 2 are available.). Using the formulae for reflections at interfaces and similar effects given by the Fresnel equation (see equation 4.3 in Chapter 4), estimate the efficiency of your modified design. How might you improve on this design? Does your design include a consideration of the glasses that are available glasses for the lens? If not, what would the implications be of replacing your 'optimum' glass with the best available?

(c) We have also mentioned lasers in the context of light generation from semiconductors and gases. The operating mechanism was described in the text. It involves a combination of a photon bouncing back into the active region and being absorbed and the consequent excited state in the material producing another, coherent, photon before it hits the electrodes in the generation region to become 'electric current'. Think through this process carefully and see if you can (perhaps collectively) arrive at a relationship between regeneration times, lifetimes and feedback percentages in the process: the photon lifetime needs to be matched by the regeneration time for the steady state to be reached. Look up on the

internet the laser 'rate equations' to rationalise your thoughts with the algebra! It can all be found in the description in the chapter, but careful thinking through the process is essential!.

5. (a) Polarising sunglasses are readly available. Try twisting the frame of a pair of polarising sunglasses through 90° on a sunny day and explain the differences you see whilst changing the angle. Then take two pairs and cross the polarisers so that no light comes through. (You may also have access to polarising sheets – readily available from many internet suppliers.) Place a transparent plastic sheet between the two crossed polarisers and stretch it in various directions. Explain the patterns and also the colours you see.

   (b) The above approach is sometimes used as a means for examining stresses in plastic samples. Usually the results are taken qualitatively – how would you make them quantitative?

6. (a) You have been given the task of designing a silicon photodetector. In silicon the saturation velocity for electrons (roughly speaking, this is the 'terminal' velocity at which any increase in field causes no further increase in velocity) can be taken as 10^7 cm/s and saturation is reached at a field level of 20 kV/cm. The ionisation field can be taken as ~300 kV/cm. Assume a 10-micron-long slightly p-doped depletion region and 1 micron of p^+ (see Figure 3.11(b)) In your estimates, use Gauss's law to arrive at the doping levels needed to achieve the field levels for saturated velocity operation from the carriers, and from this estimate obtain the bias voltage.

   (b) What would be the potential maximum detection bandwidth for modulation applied to the optical signal incident on the structure?

   (c) Look up the optical absorption coefficients in silicon and comment, with reasons, on the wavelength range over which this 10-micron / 1-micron design is practical.

4
Light Interacting with Structures

In the previous chapters we explored the essential features of the interactions between light and homogeneous materials. In fact, we may have cheated a little by already looking in Chapter 2 at the optical characteristics of a planar interface between dissimilar homogeneous materials and in Chapter 3 by discussing the device implications of structures such as photodetectors. The present chapter is dedicated to expanding this discussion into, initially, non-planar interfaces, and thereafter examining the behaviour of material structures and how this varies as the structural dimensions, compared with the optical wavelength, are progressively reduced.

4.1 Structures with Dimensions Much Greater than the Optical Wavelength

The simplest possible material structure is a straightforward planar interface between two different but transparent optical materials, as discussed in Chapter 2. In Chapter 2 we also mentioned the polarisation dependences of reflection and refraction at such interfaces. Polarisation is in general an important basic feature of electromagnetic waves – most noticeable in the optical domain but present throughout the spectrum. The concepts are explored in more detail in Appendix 1. The essence of that exploration is that polarisation states can be manipulated by birefringent components (i.e. components having polarisation-dependent optical paths) and that any given polarisation state can be split into components aligned with the principal axes (i.e. each of these optical paths) of the birefringent medium. These principal axes can relate to linear states – by far the most usual situation – or to circular states in circularly birefringent (chiral) materials.

Configurations based on birefringent transmission paths – often involving a separate component known as a phase plate – have an enormous range of application in the tuning of polarisation states in order to make specific measurements on materials, for example. In reverse, they are used in the analysis of reflected and/or transmitted polarisation states. There are many possibilities since, for example, one polarisation state may see an entirely different index behaviour (the index may go through an absorption resonance) and so can yield important information about the material through which it has passed.

Prisms and Lenses

Optical devices based upon large material structures are extremely common in everyday life, from videos screens to the focussing optics in CD players, to celestial telescopes, to raindrops forming rainbows. The last of these exemplifies one very important feature of most interfaces between two different materials, namely that the refractive indices on both sides of the interface are a function of wavelength. For most solid materials (glasses for example) in the visible the blue gets the biggest 'kick in the pants'(i.e. the largest deviation); this is often referred to as normal dispersion. The triangular prism exemplifies this (Figure 4.1).

Usually the prism is seen as a means of dispersing optical wavelengths, as indicated in the figure. Prisms, however, have other interesting properties, particularly those with a right-angled isosceles triangle cross section. A light beam incident perpendicularly on the hypotenuse of such a triangle will go straight ahead (though some will be reflected), will hit the far face at an angle greater than the critical angle and therefore will be totally internally reflected and return in the same direction but displaced by an amount determined by the prism geometry and the position of the incident beam. Reflecting prisms of this nature have extensive uses and have also emerged in slightly modified forms to

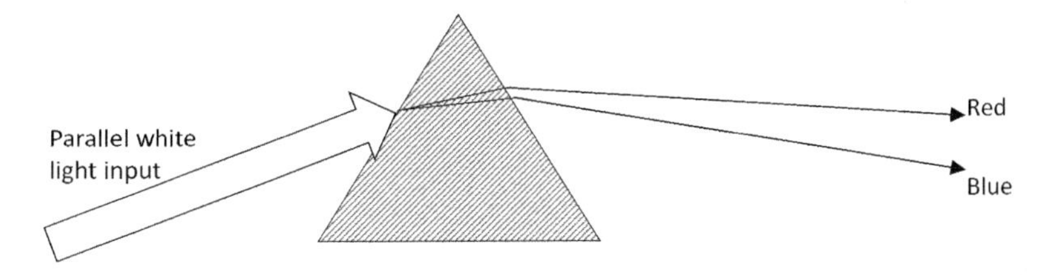

Figure 4.1 Dispersion from a triangular prism with a parallel white beam incident at an angle on one face showing the effect of the different refractive indices for red and blue.

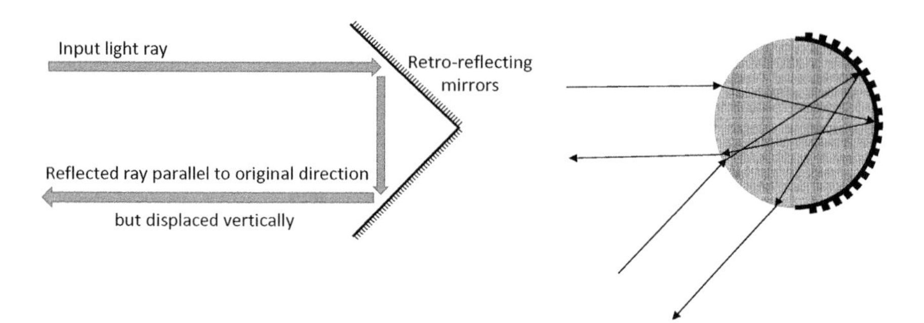

Figure 4.2 Retro-reflecting mirrors and a spherical-lens retro-reflector.

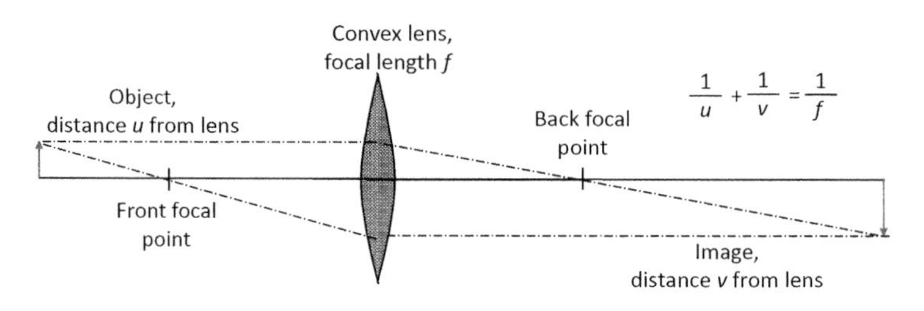

Figure 4.3 The paraxial approximation image relationships for a thin convex lens illuminated by a parallel beam.

make, for example, compact large-area reflectors and even to form the basis of the 'pentaprism' viewfinder systems in some types of single-lens reflex cameras.

These features of the right-angled prism give this direct reflection effect even for slight changes in the angle of incidence of the input beam. This back reflection effect also occurs for two mirrors with their surfaces at right angles to each other, as indicated in Figure 4.2 – a configuration often referred to as a *retro-reflector*. This retro-reflector is somewhat limited, because a more careful examination shows that it only works for an input beam incident at right angles to the out-of-plane surface of the mirror. However, this can be made into a much more versatile system: a corner cube retro-reflector, with reflecting surfaces on three mutually adjacent surfaces of an opened cube. Taken a step further, a spherical retro-reflector can effectively cope with a wide range of input angles, as also illustrated in the figure. These retro-reflectors and several variants thereon feature prolifically in everyday life, in the ubiquitous cat's eyes on roads, in reflective jackets, traffic signs, LED optics and a host of other places.

The lens – typically spherical and convex, as sketched in Figure 4.3 – is probably the most familiar imaging structure. The figure also indicates the basic principle of operation through refraction at the input and output faces.

For the so-called thin-lens *paraxial approximation*, where all the angles of incidence and refraction obey the $\sin\theta \sim \theta$ approximation, the lens formula applies:

$$\frac{1}{u}+\frac{1}{v}=\frac{1}{f} \tag{4.1}$$

where the object distance u is related to the image distance v through the focal length f. This shows that, for example, if the object is at infinity (that is, a parallel beam arrives) then the focal point is also the image point.

For a practical lens to be used in the visible range, an important caution must be noted. We have already mentioned that the refractive index for the blue part of the spectrum is typically higher in glass than that for the red part of the spectrum, so the focal length for blue light will be shorter than that for red light – giving rise to chromatic aberration. There is also another important observation – the lens is an aperture and so not only will there be ray propagation through the lens but there will be diffraction (discussed later) at the edge points; this diffraction will in turn create a 'blur' on the idealised focal point. The angular spread of this blur will be of the order of the wavelength divided by the lens aperture D and consequently the blur reduces as the aperture gets larger. This also leads to criteria for lens resolution, since the minimum feature size that this lens will be able to discriminate when imaging will also be of the order of the spot size. Again, this is wavelength dependent, with better resolution in the blue than the red.

Increasing the resolution – an incessant demand whether for cameras, telescopes or microscopes – automatically implies increasing the aperture of the lens, in other words, the ratio of the diameter to the focal length. This takes us away from the paraxial approximation, with implications shown in Figure 4.4. For a simple spherical surface, the focal length becomes a function of the distance away from the lens axis in the manner indicated. However, it is

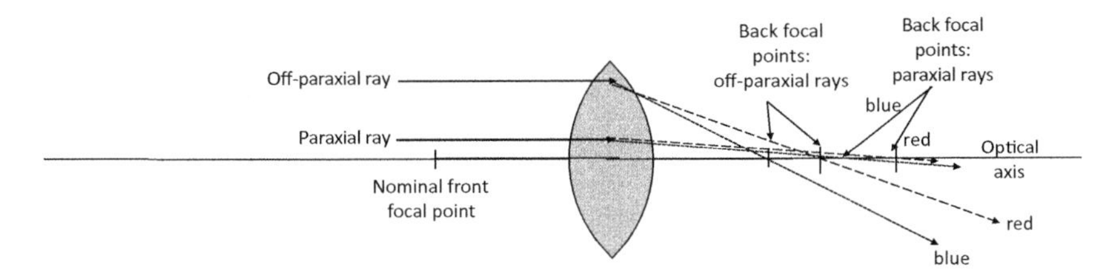

Figure 4.4 The off-axis impact for a thick convex lens, indicating changes in the focal length with wavelength and the range of foci for rays incident at varying distances from the lens axis.

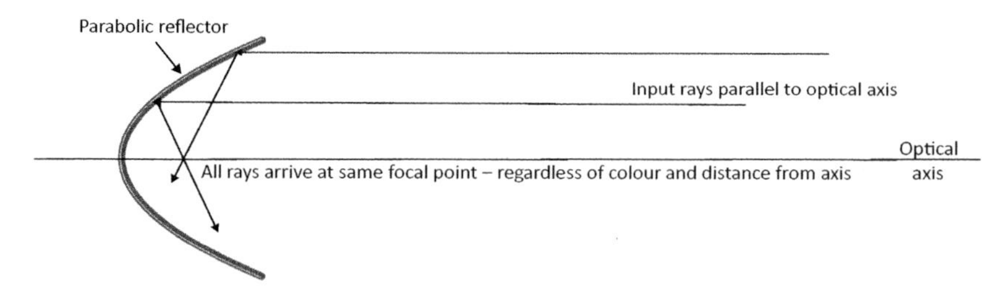

Figure 4.5 A parabolic mirror reflector, for which all rays arrive at the same focal point, regardless of their distance from the mirror's principal axis. The image receptor is at the focal point.

relatively straightforward to calculate the 'aspherical' shape necessary to ensure a constant focal point. Aspherical lenses are now commonplace but they function precisely for only one wavelength and acceptably so for a limited band of wavelengths around that particular one. All currently available glasses are chromatically dispersive. There are ingenious corrections for this, typically involving combinations of glasses with different dispersion characteristics and also with concave and convex surfaces. The details are involved and complex and require an intimate knowledge of the range of available glasses and their optical properties. There are some commercial software packages which ease the burden, but high-aperture lens design remains a specialised area.

Mirrors

The whole process becomes much simpler using mirrors (Figure 4.5). Mirrors have been particularly useful in astronomical telescopes since then almost any object of interest is, to a very good approximation, an infinite distance from the mirror so all images will cluster around the focal point. The mirror can be made large, so that a useful imaging area is available without too much masking of the input light (also, a large mirror both maximises the light collected from distant stars and galaxies and enhances stellar-image resolution). The *acceptance angle* of such a system is also limited to the order of D_i/f, where D_i is the dimension of the image receptor at the focal point, typically in centimetres, and f is the focal length of the mirror, typically in metres.

If the mirror is to provide a perfect focus, first of all its front surface must be coated with the reflecting medium, rather than its rear surface. Also, the surface must be very accurately machined and kept totally clean thereafter. Second, its shape has to be exactly parabolic to an 'optically flat' degree, implying better-than-wavelength precision! The details of this shape are given

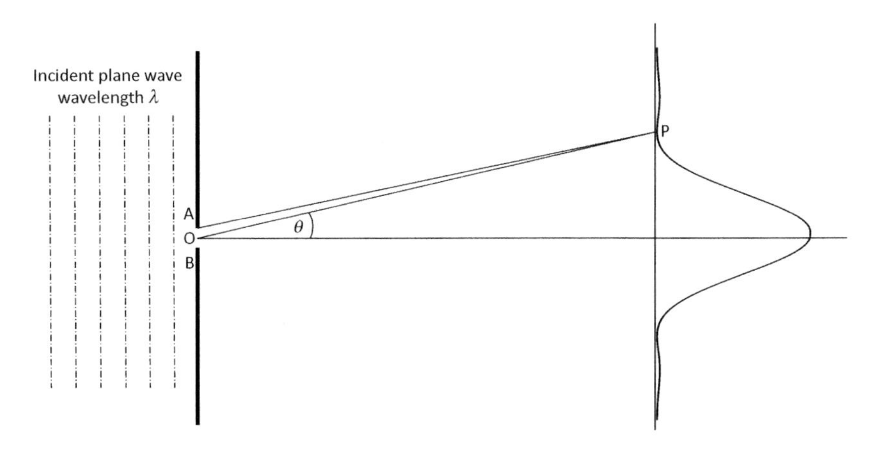

Figure 4.6 Diffraction from a single out-of-plane slit AB, of width a in a screen. The first minimum occurs when OP – AP ≃ BP – OP ≃ $\lambda/2$.

immediately by the focal length required. Since, as mentioned above, in astronomical telescopes the mirrors are typically metres in diameter, this gives an idea of the machining tolerances needed in making such a mirror. Precision grinding to the necessary parabolic shape is an essential part. Additionally, the need for extremely precise assembly throughout and very stable operating conditions (thermal expansion in the mirror materials can cause severe distortions, for example) becomes apparent.

Atmospheric turbulence also needs to be avoided. We have all seen the effect hot air rising from a black road surface. Such effects can severely limit the imaging capacity of these telescopes and hence the desirability for locating them on a high mountain or, even better, extraterrestrially. Furthermore, very little light is received from distant stars, so the image collection system needs time to acquire a workable image and the telescope must be stable with respect to the star (not the earth) as the image is collected. Hence, precision positioning systems are also mandatory, complemented by precision manoeuvring, to ensure that the stellar image remains fixed on the image receptor for the entire observation time.

Material structures which are large compared with a wavelength obviously have many manifestations in both daily and scientific life, from camera lenses to bicycle pedal reflectors through precision imaging, whether at the microscopic or galactic scale. There will also no doubt be some readers with corrective spectacle lenses!

The prism has been a central element in spectroscopic measurement systems, but recently prisms have been largely superseded by diffraction gratings – which we cover in the next section. Diffraction grating spectrometers need lenses, however. Overall, our knowledge of both the intergalactic and microscopic worlds relies upon the understanding and exploitation of the concepts which have been outlined here.

4.2 Structures with Dimensions of the Order of the Optical Wavelength

We are now concerned with structures whose dimensions are of the order of 0.1 to 10 times the wavelength of the incident light, which is just a few micrometres. Arguably, this is the most important dimensional range in the study of photonics. Even the relatively large structures which we examined briefly in the previous section inherently rely upon precision interfaces defined to this tolerance. We mentioned the precision grinding of astronomical telescope mirrors, for example.

It is interesting that domestic window panes are, thanks to the float glass process invented in the twentieth century, surprisingly flat. On the other hand, look through the surviving panes of glass from before the twentieth century and the ripples are immediately obvious. Also – in all cases – the tiniest scratch on a glass window becomes immediately apparent despite its having dimensions within the micrometer range.

So why is this tiny scratch so apparent? In simple terms the scratch blocks out the transmitted daylight, so that its presence is evident. This can of course be described just by the ray model of light. However, if in dark conditions a laser is used to illuminate the scratch, a pattern of spots due to the scratch will be seen: the laser light diffracts around the corners of the scratch and an interference pattern is formed. An interference pattern would also be formed in daylight, but it would be hard to see it.

The inverse situation to the scratch in window glass is that of a single narrow slit of width a in an opaque screen illuminated by a monochromatic light source (Figure 4.6). According to Huygens' principle, semicircular waves will diffract from each point on the slit. The waves will interfere with each other. If we travel far enough away from the slit, say a distance D_f, constructive interference will occur at certain directions. The diffracted beam centred around the first outer peaks will separate from the straight-through beam. The distance D_f is given by:

$$D_f \sim a^2/\lambda \tag{4.2}$$

(which assumes a small angle of diffraction (for which $\sin \phi \sim \phi$), corresponding to a slit more than ~ 10 wavelengths wide). This particular distance D_f is significant. It is known as the beginning of the *far-field region* for the aperture, a term used interchangeably in optics, microwaves, and acoustics or anywhere waves may propagate. This region is also sometimes referred to as the Fraunhofer region. It defines a distance from the slit or aperture beyond which the perceived diffraction pattern simply spreads out in space as if from a point source (as seen in two dimensions, in Figure 4.6).

To find the zeroes in the single-slit diffraction pattern, imagine that the slit is divided into half; then, at a certain angle to the axis, the wave travelling from the top point of the slit will be exactly out of phase with the wave travelling from the point halfway down the slit. If the angle at which this occurs is θ, then the path difference between these two waves will be $(a/2)\sin\theta$, and this will equal $\lambda/2$ since the waves are out of phase. For small values of θ we thus have $\theta = \lambda/a$. Since the same reasoning applies to each pair of waves coming from corresponding points $a/2$ apart on the slit, there is complete cancellation of all the light in the direction θ, and so zero light intensity is observed on the screen at that angle. (Dividing the slit mentally into four, six etc., the same arguments apply at angles $2\lambda/a$, $3\lambda/a$ etc., to give subsequent zeroes on the screen). The theory of single aperture diffraction is important for the lenses used in optical systems.

Diffraction Gratings

We now consider what happens when there is more than one slit. The case of two slits was mentioned in Section 2.3, see Figure 2.11, and the expression for the maxima was given by $\sin\theta = n\lambda/d$. Now let us think about a device which has not two but many evenly spaced slits. Two simple diffraction gratings are shown schematically in Figure 4.7. Here we see, in an opaque screen, rulings, servings as slits, with very sharp edges which scatter the light in all directions. As for two slits, complete constructive interference occurs every time the path difference between light diffracted from adjacent slits is an integral number of wavelengths. For wavelength λ and slit separation d, this again occurs at angles θ_n given by

$$\sin\theta_n = \frac{n\lambda}{d} \tag{4.3}$$

Since we have assumed that the diffracted light is equal in intensity in all directions, then all these image lines, at angles θ_n, would be of equal intensity. In the region where $\sin\theta \sim \theta$ the angle θ is related linearly to the inverse spacing, $1/d$, of the grating (i.e., the spatial frequency per unit length of the grating) and the images are all of equal intensity. For clarity, only the first-order angle of diffraction for each structure is shown in the diagram. The central observation here is that the angular dispersion of a grating near the axis,

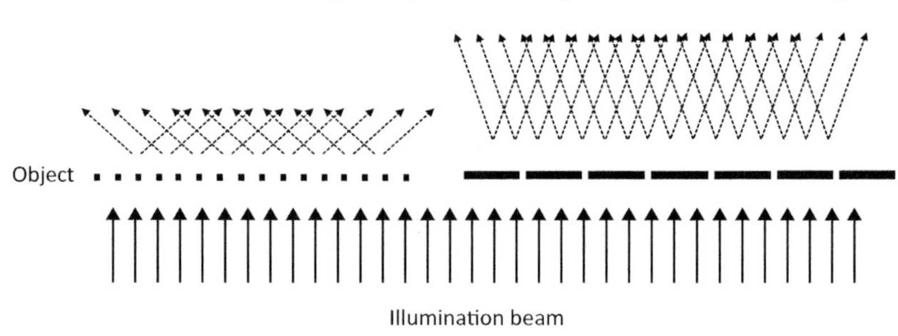

Figure 4.7 The finer the detail in an object the wider the angle of diffraction and therefore the wider the aperture of the lens required to collect that detail.

where $\sin\theta \sim \theta$, is in fact a *Fourier transform* of the spatial grating pattern; here it is a good approximation to a row of periodic delta functions. This is not a coincidence, since it is relatively straightforward to demonstrate that in the paraxial region this Fourier transform relationship between any structure such as a grating and its far field is universally applicable. (This is discussed further in Appendix 2.) This Fourier transform relationship is an extremely powerful tool, used in many situations in photonics.

Thus far, we have assumed that the grating has infinite extent (aperture), but of course it is not actually infinite, so what about the edges? First of all, consider an aperture similar to that indicated in Figure 4.6 – a wide slit aligned vertically to the plane of the diagram. Illuminating the aperture corresponds to a single simple pulse and its Fourier transform is a sinc function. Then consider the spacing of the zeroes in the sinc function. Again, for small diffraction angles, as we saw easlier the zeroes are given by $\theta = m\lambda/a$, where $m = 1, 2, ...$ and a is the aperture (width) of the slit.

Suppose this aperture is now used in front of an infinite square wave grating. In effect we are multiplying the image spaces together, so consequently we will get a convolution in Fourier transform space, as indicated in Figure 4.8.

This relationship has many implications, of which perhaps the most obvious lies in the use of a microscope as an imaging system. The general principle here is that the object is placed in the focal plane of the objective lens to produce eventually an image at infinity in the eyepiece. In order to resolve the fine detail in the object the imaging system has to collect as much of the light diffracted from the object as possible. Consequently, the larger the ratio of the diameter of the input objective lens to its focal length, the more detail it can collect and the higher the resolution of the microscope. This is another facet of the 'blurring' due to finite aperture size mentioned earlier.

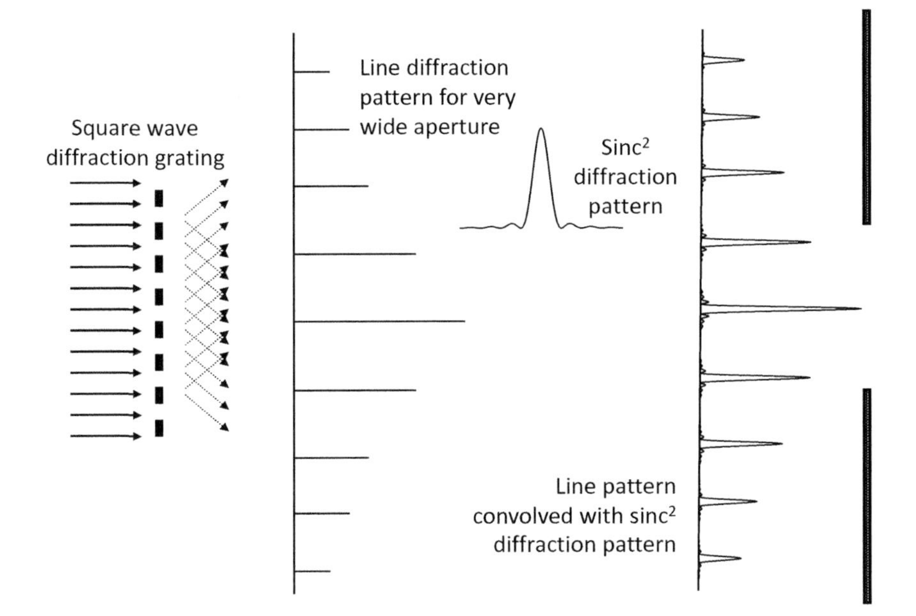

Figure 4.8 The diffraction pattern from a square wave diffraction grating and its modification after passing through a finite but wide aperture (see text). Also indicated, on the right, is a spatial filter, a narrow aperture allowing through only the central three diffraction orders – what would be the image following this spatial filter?

These observations also point toward techniques for image processing and image enhancement, much of which take place by manipulation of the spatial frequency content of the image using optical elements (or even software tools on digital images) in the so-called Fourier transform plane (Figure 4.9). This is in the focal plane of the imaging objective; here the diffracted beams from the object are focussed to produce an image of the spatial frequency spectrum (strictly speaking, a squared image, since what we see is intensity rather than amplitude). A basic example was given in Figure 4.8. Here we started with a square wave grating and producing the Fourier transform distribution as indicated. Each of the image lines has an amplitude sinc function around it arising from the finite aperture of the lens and/or grating. Suppose we then insert a spatial filter (narrow aperture), shown in Figure 4.8, which allows through only three beams, corresponding to the zeroth-order and first-order diffracted beams. The image no longer fully represents the original. And what would the resulting image be if this spatial filter is placed in the front focal plane of a second lens? It can be shown that if on the other hand we attenuate the central (low spatial

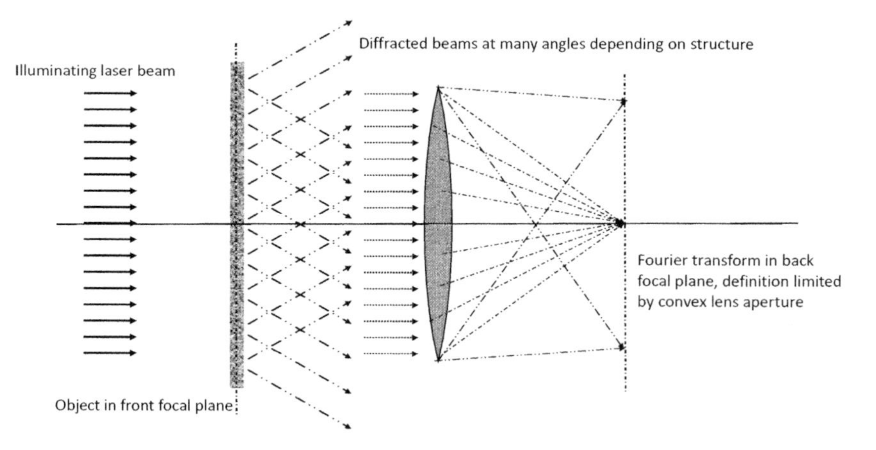

Figure 4.9 Collecting the diffraction pattern from an object in the front focal plane of a lens produces a focussed representation of the periodic components as the image in the back focal plane. Also, note the impact of the lens aperture, which cuts off the higher angles of the diffraction pattern.

frequency) beam with respect to the higher-order beams then the result is edge enhancement – a sharpening of the image definition. Spatial filtering, by means of aperture, attenuating screens, or even spatially varying phase delays (see below), is a frequently used tool in image enhancement.

The Fourier transform relationship between an object and its far-field diffraction pattern (or between the input plane and the transform plane in a simple convex lens system) also applies to transparent objects, many of which have structure within them which is invisible in a conventional image. However, this structure will correspond to variations in optical thickness – that is, optical phase and phase functions also have Fourier transform properties. For example, a sine wave variation in the optical thickness of an object (which corresponds to a spatially varying phase delay) gives rise to a set of sideband frequencies which recombine to produce a zero-contrast image, with no intensity variations. In the simplest case we can remove the central background component – the average transmission through the object – to obtain the frequency-doubled intensity distribution indicated in Figure 4.10(a). Whilst this is distorted, it can be interpreted by an experienced observer. This technique is known as dark-field imaging. You may like to consider why in Figure 4.10(a) the image is spatial-frequency doubled, as indicated, with a zero every half-period of the original sinusoidal phase grating.

There is also a related technique, known as the phase contrast method, through which the phase of the zero-frequency component is shifted by 90° (this will only apply exactly for monochromatic illumination. Useful results

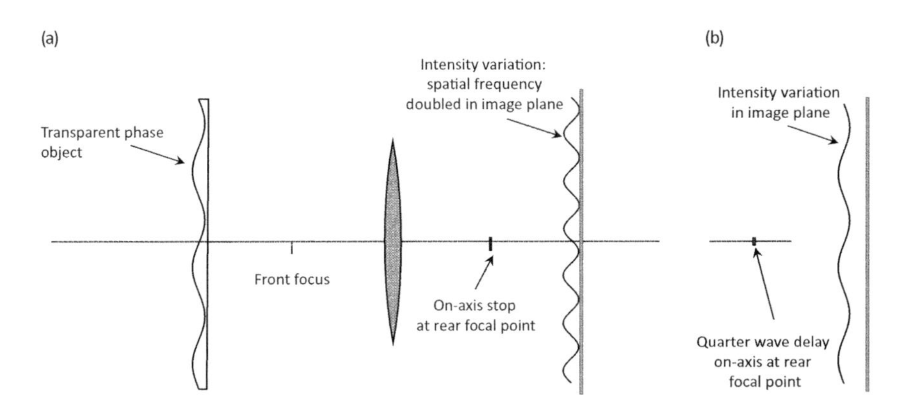

Figure 4.10 Making phase changes in a transparent object visible: (a) using an on-axis stop to produce a frequency-doubled intensity image or (b) using an on-axis quarter-wave delay plate to produce an intensity image of the original phase distribution.

can, however, be obtained over a relatively wide range of wavelengths – the reader might like to estimate what this range might be). This technique can also, for small phase deviations, produce an amplitude image which is directly related to the original 'phase object'.

Three-Dimensional Imaging

The above brief discussion contains much conceptual food for thought. It also points towards the principles of holography – in effect three-dimensional imaging without lenses. Here light scattered from the object (Figure 4.11) again forms a diffraction pattern but in this case the diffraction pattern is combined with a larger sinusoidal phase-reference beam extracted from the same illuminating laser. The interference between the diffraction pattern and the laser beam produces an intensity distribution containing sinusoidal components exactly related to the diffraction angles, and therefore containing all the three-dimensional information about the illuminated areas of the original object. Consequently, recording this pattern to make a permanent hologram and subsequently shining a laser onto this modified diffraction pattern retraces the original paths and the viewer sees a three-dimensional image. Need this laser be the same wavelength, and why do we need the high-intensity reference?

Imaging is among the most important manifestations of photonics. This section has highlighted most of the essential underlying principles of imaging systems. Some demonstration experiments with everyday objects – a magnifying glass, a laser pointer and possibly some simple samples and apertures

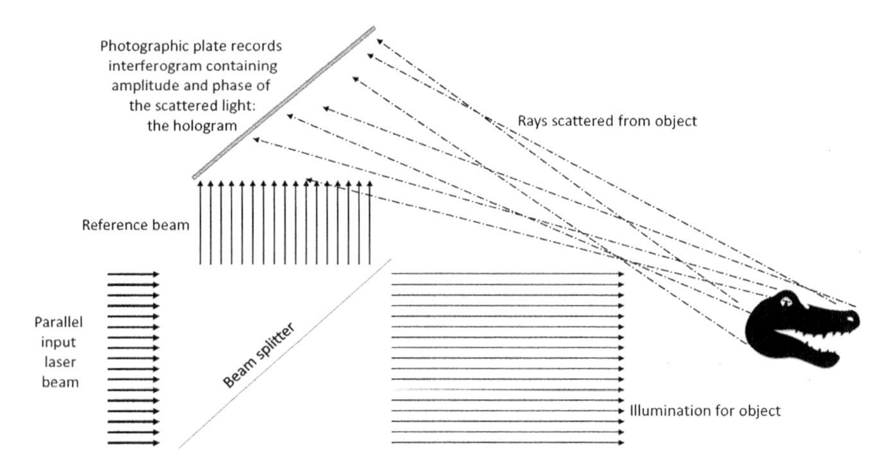

Figure 4.11 Making a hologram on a photographic plate, shown at top left of the diagram.

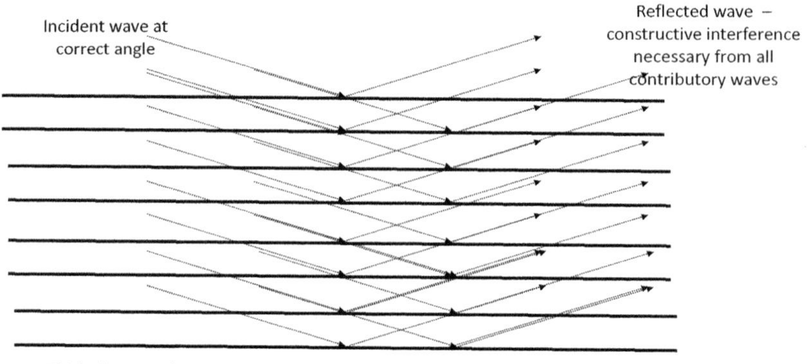

Figure 4.12 The essential concepts of Bragg diffraction. A three-dimensional grating is formed by a set of crystal planes on by a high-frequency acoustic wave, which forms a periodic variation in refractive index.

which can illustrate these principles – are outlined in the problems at the end of the chapter.

In preceding discussion we implicitly assumed that all the diffracting objects were thin compared with the wavelength of the light which is being diffracted. Thick gratings – phase objects many optical wavelengths in thickness – also play their part; this is exemplified in Bragg diffraction (Figure 4.12). In optics the Bragg grating is typically a phase object structure which can be created by, for example, the pressure variations induced by a travelling ultrasonic wave.

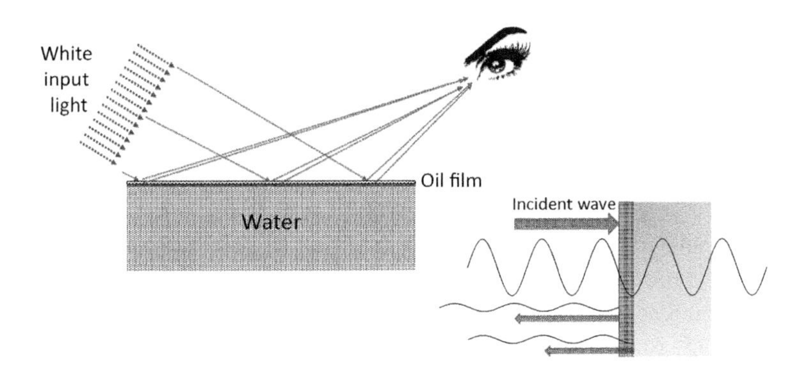

Figure 4.13 (left) The oil-on-water effect embodying multiple reflections from the surface and the oil–water interface and (right) the incident and reflected beams for normal incidence.

For an incident beam at the appropriate angle and an appropriate total phase delay within an ultrasonic-wave grating, the entirety of the incident optical beam can be diffracted as indicated in the diagram. Since the 'grating' is moving, the diffracted beam is Doppler shifted in frequency by an amount identical to the input frequency of the ultrasonic driving signal, a feature which is often used in optical frequency shifters.

Finally, another observation on diffraction gratings. Suppose a parallel beam of white light, rather than monochromatic light, shines upon a diffraction grating of the type shown in Figure 4.10. In this case the blue component will be diffracted rather less (by a factor of 2 or so) than the red component; therefore, by appropriate spatial filtering, the apparent colour of this grating can be made to change – the first hint that perceived colours depend not only on spectroscopic properties, as discussed in Chapter 3, but also on the structural properties of both a material itself and the system through which it is observed.

Thin Films

A commonly observed example of this is the coloured structure of a film of oil on water. Figure 4.13 illustrates how this might happen. Oil and water have different refractive indices and therefore both the interface between the oil and the water and the interface between the oil and the air will introduce partial reflection, with amplitude reflection coefficient R. At each interface, at normal incidence R is given by the *Fresnel equation*, namely

$$R = \frac{n_1 - n_2}{n_1 + n_2} \tag{4.4}$$

where n_1 is the refractive index of the medium from which the light is incident on the interface and n_2 is the refractive index of the medium into which it passes (this equation is a special case of the Fresnel reflection relationships mentioned in Appendix 6.2). Note that this equation 4.3 implies a phase change of 180° for light incident from a low-index material to a higher-index material. The value of the reflection coefficient R (remember that it is an amplitude), for light travelling from air to glass, where the index of the latter is 1.5, is around 0.2, corresponding to 4% power reflection. The situation for the case of oil and water is slightly different, in that a typical oil has an index of approximately 1.4 whereas water sits at 1.33. Consequently, there is no phase inversion at the oil-to-water interface and, typically, the amplitude reflection coefficient for normal incidence on the oil-to-air interface is about 0.14 compared with –0.1 (the negative sign indicates 180° phase reversal) at the water-to-oil interface. However, reflections from the two interfaces will interfere, with a range of possible results varying from constructive interference, producing a total net reflected amplitude of 0.24, to destructive interference, producing 0.04. This is a 30 to 1 ratio in perceived optical intensity (we have ignored multiple reflections since these are relatively small though they will have some impact). The visible spectrum covers around a factor of 2 in wavelengths so the above discussion indicates that, depending upon the oil thickness and the angle of observation, the reflected colour will vary throughout the spectrum – another example of where the spectroscopically defined colour of a material can be radically changed by its structure.

Thin films and the design and realisation thereof constitute a vast topic. Antireflection coatings are a common application. If we arrange for the two reflected beams to be out of phase (this happens if the film thickness is one quarter of a wavelength *and* the reflection amplitudes are identical for each interface) then the resulting destructive interference at the surface implies zero reflectance. This concept finds widespread applications. Camera lenses, spectacle lenses and a host of other optical surfaces typically exhibit a bluish tinge when viewed from an appropriate angle. This is a manifestation of anti-reflection coatings. These are incorporated to minimise the impact of reflections on image quality typified by the glowing, often yellowish, circles seen in photographs taken when the camera is pointed towards, but not directly at, the sun.

Waveguides

Figure 4.14 shows the simplest possible implementation of an optical waveguide, a step-index fibre. Here, provided the light is incident on the interface between the core and cladding at an angle exceeding the critical angle, the light will be guided through the higher-index core region. Typically, light would be launched from some external source – often air. There will be a maximum

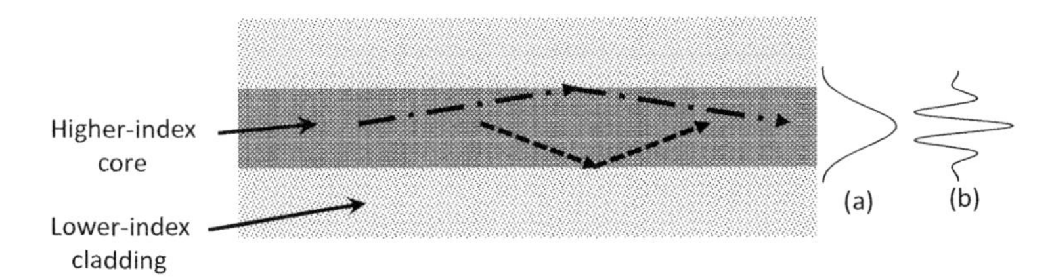

Figure 4.14 A step-index optical waveguide showing, in the main part of the figure, ray paths for two different modes of propagation and, on the right, the light amplitude distributions for (a) the lowest-order mode and (b) a slightly higher mode, corresponding to interference between the beams at steeper angles of incidence. See Appendix 4, Figure A4.2.

angle that the incident launched light beam can make with the axis of the waveguide, beyond which guiding cannot take place. This angle corresponds to the critical angle for light travelling from the core to the cladding. The *acceptance angle* θ_{na} is given by

$$\sin\theta_{\mathrm{na}} = (n_{\mathrm{core}}{}^{2} - n_{\mathrm{clad}}{}^{2})^{1/2} \sim (2n\Delta n)^{1/2} \tag{4.5}$$

where n is the average refractive index of the core and cladding. The quantity $\sin\theta_{\mathrm{na}}$ is often referred to as the numerical aperture.

The approximation in equation 4.4 is often applicable since the index difference Δn is typically small compared with the indices themselves. (It is straightforward to verify this from Snell's law relating the on angle of incidence and the angle of refraction.) This basic structure forms the basis of the optical fibres used to interconnect domestic electronic appliances or to provide short-distance low-capacity communication links.

One limitation of these optical waveguides becomes apparent when one observes that there are two extreme paths which the light rays can take and a whole host of other paths in between: the path difference between a central-axis ray (the lowest-order mode) and a ray reflected at the critical angle can be estimated as $l\Delta n$, where l is the waveguide length. For index differences of 1% and time delays of 5 μs per km, which are typical, this path difference corresponds to ~50 ns of dispersion over a 1 km length of waveguide. For monochromatic light this dispersion is purely an artefact of the structure of the waveguide. Thus pulses of 20 ns duration (about 20 Mbits/s – slower than many domestic broadband speeds) will be smeared out completely within less than a kilometre and therefore lose all value. However, these step-index fibres are very simple to produce and can be made from plastics as well as glass, and so for short distances they find extensive application.

Perhaps one way to solve the problem of dispersion is to design a structure which has only one propagation path, leading to the concept of the single-mode

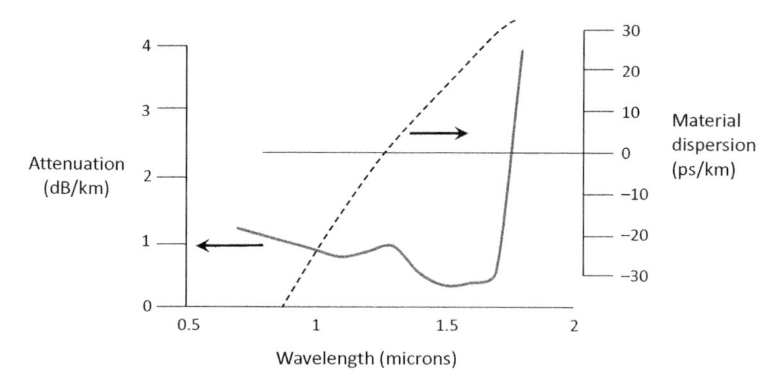

Figure 4.15 Losses (solid line) and material dispersion (broken line) in silica. The zero dispersion point at 1.3 μm is a small, but significant distance away from the lowest attenuation at 1.5 μm wavelength. See Section 5.7.

optical fibre. These fibres have been extensively used for over three decades in long-distance optical fibre communications and among other things are essential in fuelling the even-increasing demand for domestic and industrial bandwidth. The internet could not function without optical fibres, as one of many vital contributing technologies. Even the single-mode fibre, though, has its inherent limitations. There is dispersion of different wavelengths in the glass which forms the fibre as well as the dispersion introduced by the optical structure; longer wavelengths penetrate further into the cladding via the evanescent wave (see below). Furthermore, there are spectroscopic wavelength-dependent losses within the materials themselves. The losses and the dispersion characteristics of silica (the principal raw material for optical fibres) are shown in Figure 4.15.

The basic operating mechanism of the single-mode fibre points towards a wave rather than a ray interpretation of transmission along a high-index region surrounded by lower-index regions. In this wave interpretation, only a limited number of ray directions are permitted even in the large-scale case of Figure 4.14. The reason is that across the interface between the high-index region and the lower-index outer region there must be continuity in the electric field. In the case of a perfect metallic interface this would imply that the electric field is zero at the interface but at a dielectric interface the continuity is a little more complex and requires some limited penetration of the electric field into the low-index region; this is often referred to as the *evanescent tail.*The net result is a field distribution of the type shown in Figure 4.14(b). For the two-dimensional case (in other words, a film enclosed between two other films), this can be considered as interference between two beams at a specific angle (Figure 4.14 (a)). In principle there are other options which would fulfil the same basic criteria, namely that at the boundary between the dielectric interfaces the appropriate field relationships hold. An example of such a higher-order mode is also indicated in Figure 4.14(b). This mode has more than the minimum half wavelength of an interference pattern across the core region. However, in order to achieve this second interference pattern, the imaginary beams which are interfering need to be incident on the interface at a higher angle. For a single-mode fibre, this angle would be chosen to exceed the critical angle and therefore the higher-order mode would not be guided. Hence a combination of index difference and core dimensions gives rise to the modal structure of the fibre. The circular cross sectional structure characterising an actual fibre involves slightly more complex beam structures but the underlying concepts are identical. The basic idea of viewing guided waves as interfering beams for which the interference pattern satisfies the necessary boundary conditions along the wave-guiding interface gives a very useful insight. Essential features such as the ideas of spatial waveguide modes, the number of modes possible in a given structure and important propagation features like intramode and inter-mode dispersion can all be approached in this manner.

The waveguide structure has other interesting implications. There is obviously a range of wavelengths for which the basic single-mode boundary conditions can be fulfilled and indeed, in principle, this range extends well beyond the optical region to the microwave and radio spectrum, though the strength of the waveguided beam becomes vanishingly small at longer wavelengths (less and less of the energy is held in the core region and eventually, at wavelengths significantly longer than the structural cross sectional dimensions, most spills out into the surrounding air). At shorter wavelengths there comes a point where at least two angular options can give rise to an appropriate interference pattern and therefore satisfy the dielectric boundary conditions, a point at which the waveguide becomes over-moded.

In the single-mode-transmission case there are dispersion mechanisms at work similar to those on which we commented for multimode guides – namely that the variation in reflecting angle for the imaginary interfering beams will inherently give a variation in propagation time, with shorter wavelengths experiencing longer delays due to geometric dispersion than the longer wavelengths – a similar trend to that experienced due to material dispersion in the 'normal' dispersion region, of a dielectric.

These dispersion effects are significant and have stimulated much ingenuity in refining the basic concept in order to cram more and more terabytes along a single fibre. There are other factors which limit fibre transmission distances, of which the most important are the inherent losses in the silica which forms the basis of the core of the fibre. These become very low (<1 dB/km) in the 1.5 μm

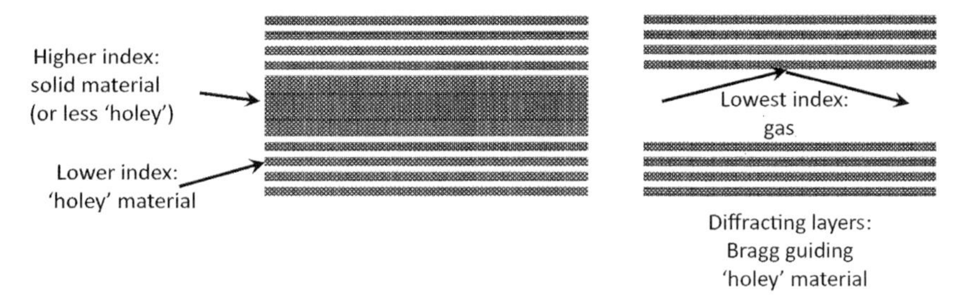

Figure 4.16 Two examples of simple 'photonic crystal' structures, showing how structures can provide innovative waveguide geometries. The structure dominates the optical transmission properties of the material from which it is composed.

region (Figure 4.15). The material itself has another interesting property: its inherent dispersion goes through zero in the 1.3 μm region – a case in which nature conspires to be helpful to the technologist. However, the dispersion in the silica at 1.5 microns, the lowest-loss region, remains significant, even at ~20 ps/km, for very-high-capacity communication systems.

Consequently, much has been achieved to minimise dispersion in this 1.5 micron wavelength region through the ingenious use of structural and material properties, so that the transmission of hundreds of gigabits per second, and more, over hundreds of kilometres through a single channel is now available.

The losses in silica exemplify different basic phenomena. The loss at long wavelengths in the infrared is fundamental to the absorption bands within silica itself and consequently there are many practical difficulties in extending optical fibre transmission much beyond the 2 μm wavelength barrier. The attenuation glitch at about 1.3 μm is due to water absorption but, thanks to a great deal of processing effort, this can be virtually eliminated. At shorter wavelengths the impact is seen of tiny inhomogeneities within the structure itself and those that inevitably arise in the processing and preparation of the fibre owing to the slightly random process of cooling down. These tiny inhomogeneities cause Rayleigh scattering which technically applies to structures much less than the wavelength; we shall return to this in the next section.

The increasing precision of fabrication technologies has continually expanded the level of control available for these photonic materials. A conceptual example is shown in Figure 4.16. Here – on the left – a number of voids (holes) have been introduced into the structure of an optical waveguide and these voids will in turn reduce the local effective refractive index, in the planar case with different changes for vertically and horizontally polarised inputs. The key point here is that the holes are a small fraction of a wavelength in diameter, so that, as seen by the propagating wave they are effectively averaged out. This gives a birefringent guide, from which, conceivably, a structure could be found which will guide one polarisation and not the other. Indeed, as we shall see such, structures based on metallic films can be designed to do exactly that.

There are many variations on this basic theme of which perhaps the most extreme is the waveguide with no core (or strictly an air core), indicated on the right Figure 4.16. The waveguiding properties here are determined by diffraction from the structure which surrounds the core and, for a given wavelength to be transmitted both the Bragg diffraction conditions which we discussed earlier and the waveguiding material boundary conditions need to be simultaneously met. Consequently, the propagation characteristics of, for example, a circularly symmetric version of this structure – "hollow-core photonic crystal fibre" – exhibit quite complex structure including some forbidden bands, for which the Bragg condition cannot be met. Viewing this as normal index guiding is not an option!

The flexibility in design options for waveguides based on photonic crystals is enormous, featuring variations in waveguide dispersion and propagation spectra which are extremely diverse. The hollow-core fibres also have the interesting property that gases and liquids can be introduced into the void, thereby offering a new tool for spectroscopic investigation; this comes with the proviso that the core is typically less than 10 μm in diameter, so that it takes some time for fluid to enter the void and to leave thereafter. However, similar functional flexibility is available for synthesising material properties on surfaces, particularly in the planar waveguide context, thereby enabling a surface-based interaction with an optical transmission channel.

The interfaces between voids and solid materials in these photonic crystal structures are never totally smooth, so that Rayleigh scattering at irregularities (see below) in these interfaces becomes a persistent issue. Consequently, the losses within photonic crystal fibres are inevitably much higher than in the in all-solid-state forerunners, which are fabricated with very smooth glass-to-glass interface structures.

4.3 Structures with Dimensions Much Smaller than the Optical Wavelength

Subwavelength structures feature strongly in our daily lives. Thanks to the scattering from such structures, sunlight diffuses throughout a room and the sky in daylight hours is blue. Such phenomena have been reasonably well understood since around 1870. Essentially, for scattering particles with dimensions much less than the wavelength of the light, the strength of the scattered signal is inversely proportional to the fourth power of the wavelength. The scattering sites can be tiny bubbles in a glass or even minute variations in local refractive index, minute particles of ice in the atmosphere or tiny undulations in the surface of a solid. The inverse fourth power relationship is at its most evident in the blue sky. Clearly, blue light will scatter about 16 times as much as red light on its way from the upper atmosphere to the earth's surface and consequently the sky appears blue. The red sky at sunrise and sunset sounds paradoxical but a brief investigation into its origins soon provides the explanation.

Scattering is also the cause of the rapid onset with shortening wavelength of attenuation in optical fibres, mentioned briefly in the previous section.

In nature, colour originating in structural artefacts rather than in material properties is extremely common, familiar examples being butterfly wings and peacock feathers. The range of wavelength-filtering structures is immense, embracing diffraction gratings, tiny coloured mirrors, photonic crystals and nanoscale particles.

This understanding of nature's very tiny structures, and the many optical properties which they can have, has stimulated technologically based endeavours to emulate and exploit these structures. The technology, of course, needs to be compatible with structural fabrication to nanometre tolerances and in turn has become known as nanophotonics. This builds upon the accumulated experience of the past century or more in making controlled-response, significantly subwavelength, electrically functional material structures, better known as electrical circuits. As with any extension of the frequency range for electrical circuits, two basic features now need to come together. The first is a thorough understanding of the characteristics and impact of what, in electrical circuit terms, are referred to as stray components (capacitance and inductance in wires, for example); this understanding become increasingly complex as the frequency goes up. The second is the necessary technology to realise circuits with acceptable spatial and material composition tolerances. Indeed, nanophotonics has been described as simply realising electrical circuits at extremely frequencies.

Perhaps the most intriguing aspect of these synthetic 'metamaterials' is their theoretical (but to some extent demonstrated) optical properties, of which can be totally controlled by the structure itself and the materials therein. Possibly the invisibility cloak gains the most headlines! This is suggested as a material coating that redirects input light so that the light emerges apparently undeflected: the object in question is no longer visible. The detailed design of such cloaks is complex but the essential concept lies in the realisation of negative-index metamaterials (or, to be more precise, material structures), for which the refracted ray lies on the same side of the normal as the incident ray,

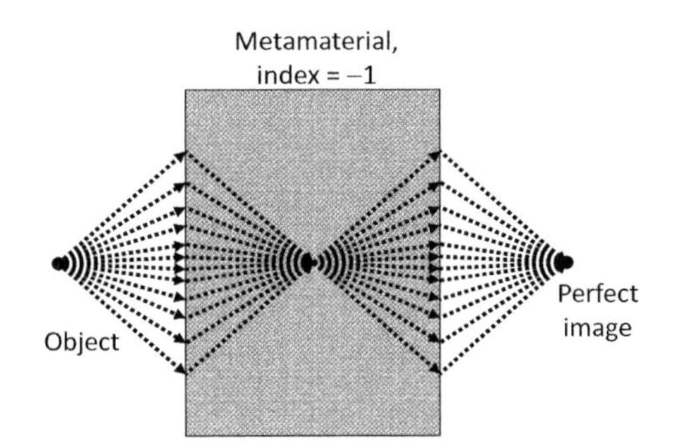

Figure 4.17 A conceptual diagram of the operation of a perfect superlens with $n = -1$.

giving rise to a negative sign in the refractive index. The behaviour of such a metamaterial structure is indicated in Figure 4.17. In principle such a structure can produce a totally undistorted image of an object. It can beat the diffraction limit, the 'blur' from the edges of the lens or mirror, which characterises traditional imaging.

The detailed principles are quite complex but some insight into the basic ideas can be derived from the observation that negative-index metamaterials can be synthesised using periodically spaced arrays of resonant circuits. A feature of such a circuit is that at low frequencies, well below resonance, the current (i.e. the magnetic field) and voltage work in phase whereas at higher frequencies, well above resonance, they are in antiphase. Within an electromagnetic wave this will attempt to pull the electric and magnetic fields out of phase with each other, a feature which can be compensated by the wave's moving to the opposite side of the normal. Metamaterials, in particular those with a negative index, have found application in acoustics (in inverting the phase of the pressure and velocity components of the acoustic wave) and in electromagnetics. Early demonstrations were at relatively low (microwave) frequencies but there have been ventures into the optical, constrained to extent by the properties of conductors at optical frequencies.

One of the principal prospects for nanostructured materials lies in the use of metallic conductors at frequencies well below the plasma resonance and at thicknesses sufficiently low for optical signals to cope with the inherent losses. One manifestation of such configurations is that the presence of a metal on an interface between two dielectric materials can significantly enhance the magnitude of the electric field on the face of the dielectric interface opposite to the direction of incidence (Figure 4.18). There is another important aspect to this. The losses are low if the electric field is perpendicular to the plane of the

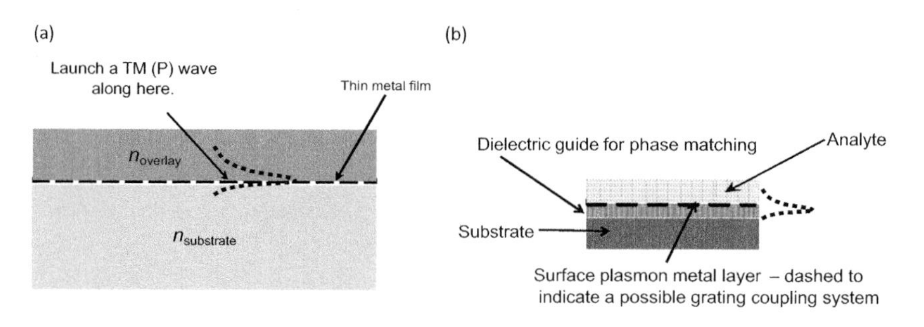

Figure 4.18 Surface plasmon waves giving high field-penetration into the overlay but requiring careful phase matching whether in multiple layers (a) or waveguide format (b). The latter arrangement finds application in, for example, liquid characterisation, hence the reference to 'analyte' (a substance under chemical analysis).

interface. However, if the electric field is in-plane, then it sees a long length of metallic material – the in-plane polarisation is very lossy. So here we have a polariser, much used over the past few decades! This basic phenomenon, which is an essential aspect of plasmonics, raises prospects for a range of photonic devices emulating their electromagnetic forerunners, from wire-based antenna arrays capable of refocussing a diverging beam to optically excited electro-chemical sensing.

Quantum Dots

Nanostructures other than butterfly wings have also been with us since ancient times. Colloidal gold nanoparticles have given permanent colour to glassware and pottery for centuries. These tiny gold spheres are perceived from red to blue over the entire visible spectrum – completely at odds with the visual appearance of bulk gold. These particles, which are typically a few tens of nanometres in diameter, have now extended their presence beyond ancient pottery into medicinal potions and (bio)chemical assay. How does this structure so efficiently modify the bulk reflected colour? There is a clue in that the larger particles appear more to be blue than the smaller particles, which veer towards the red. It would seem, then, that the large particles absorb the red end of the spectrum and vice versa for the smaller particles. Clearly, then the key lies in the dimension of the spheres but also in the observations on surface currents discussed above. A large sphere, which is typically about 100 nm in diameter, resonates in the red and therefore absorbs the red whereas a small sphere, around 30 nm in diameter, resonates in the blue.

Venturing towards even smaller dimensions – controlling them to nm or greater precision – can tailor the apparent colour of most solid materials. A large 'quantum dot' has dimensions of around 5 nm. At this 'large' size the dots tend to absorb, and equally can fluoresce, in the red. This potential for fluorescence distinguishes the behaviour of quantum dots from that of gold nanoparticles and their surface plasmon resonances. The smaller quantum dots, 2 nm or thereabouts, by the same token tend to absorb and fluoresce in the blue. In this case it is not surface plasmon waves and the electromagnetic absorption therein which is the source of the colour change for these tiniest of particles: the physical size of the structure is now so small that the quantum mechanical energy levels within the structure are involved. Consequently, the absorption band composition and hence the colour is affected. Quantum dots typically find application in the context of lasers, solar cells and light emitting diodes but also in some forms of medical imaging and in quantum computing. There are many possible applications, but most will take some time into the future to mature.

4.4 Summary

Perhaps the principal message from this chapter is that structures are an extremely flexible, even fundamental, tool for the manipulation of light, whether regarded as rays, waves or photons. The operational functionality of this tool changes in a very versatile manner with the dimensions of the structure in comparison with the wavelength of light.

For a very large structure – essentially a fraction of a millimetre dimension and above – we have seen how the shape of interfaces and of optical components, typically glasses, modifies optical functions, prisms and lenses being the dominant systems. We have also seen that using a mixture of glasses and shape functions can enhance functionality; this is most apparent in the design of compound lens systems. The basic phenomenon of refraction, together with the associated polarisation-dependent reflection effects at interfaces, is at the heart of the familiar optical components within this dimensional range.

Even such large structures are, however, affected by the impact of phenomena most conveniently thought of as relevant to the medium-dimensional range; that is, around the optical wavelength. The resolving power of lenses is the most obvious feature, as it can be explained in terms of the essential medium-scale concepts of diffraction and interference. These medium-scale features underlie holograms, fibre optics, the appearance of oil on water and many other familiar optical phenomena.

Advancing to the minute scale, we have structures with features small compared with an optical wavelength. Here we enter the domain of nanophotonics, where carefully considered extrapolations from the concepts of electrical circuits come into play. The ideas underlying plasmonics are possibly the closest conceptually to circuits. Metamaterials, another essential concept from the tiny minute scale, involve the fabrication of carefully organised material mixtures with features well below the optical wavelength, again conceptually familiar from, for example, a TV antenna.

Going to even more minute structures, of the order now of 1% of the optical wavelength on a few nm, we reach the quantum dot regime where the dimensions of the structure begin to impact upon the quantum mechanical properties of the material. Very tiny structures will, for example, shift the energy level distribution. This entirely modifies the actual optical properties of the original material. This is a relatively new domain, but it has already demonstrated for example, the potential for very flexible laser design and innovative features in solar cells.

Optical materials and material structures combine to offer great flexibility in the functionality of optical artefacts. In fact, almost every application of optical systems relies on carefully and critically combining the optical material properties as briefly considered in Chapter 3 with versatile application of structural features outlined in the present chapter.

4.5 Problems

1. (a) Think about the basic spherical biconvex lens of the type indicated in Figure 4.4. Now use your knowledge of the slopes of the spherical surfaces (just consider two dimensions) to gain some insight into how the focal point might vary as a function of lens aperture (compared with the focal length) at a single wavelength. At what aperture would this variation in the position of the focal point begin to compromise the resolution of the lens for illumination at 500 nm wavelength (think of the implications of Figure 4.6)?
   (b) It is possible to achieve a perfect (monochromatic) focus with an aspherical surface. How would you approach determining the shape of this surface?
2. This problem concerns retro-reflectors.
   (a) Consider a prism with a right-angled isosceles-riangle cross section. It is simple to demonstrate that a beam normally incident on the hypotenuse

will be retro-reflected. Over what angular range (for angles lying within the plane along the hypotenuse and perpendicular to the input beam) will a parallel incident beam be directly reflected (take the refractive index as 1.5)?

(b) How will the reflected beam angle change with incident angle? You will probably have instinctively approached the problem in two dimensions – varying the angle about an axis which is perpendicular to the hypotenuse and in the plane of the paper on which you've sketched the system. How might this change if the beam angle were varied around an axis a plane perpendicular to the hypotenuse?

3. Red sky at sunrise and sunset, blue skies in day time. Why are clouds are essential for an impressive sunset? On the same theme – why are clouds white (except at sunrise and sunset) when relatively thin, but go through shades of grey into a threatening black?
4. (a) Convince yourselves that a parabolic mirror gives a perfect focus at all light input wavelengths (Figure 4.5).

   (b) Assume that this applies to a 5 m diameter reflecting telescope. Give a reasonable estimate for the angular resolution and practical angular coverage of such an instrument. Atmospheric turbulence can distort images for even a perfect reflector. How would you approach setting up a system to overcome these disturbances?
5. (a) How would you measure the number of threads per centimetre in a piece of cloth with just a laser pointer and a piece of white paper? Do the experiment, comment on how the answers change when the cloth is stretched in one direction and try shining the pointer on a piece of scratched or smeared glass and looking at the spots that are transmitted.

   (b) What might be some practical implication of these observations in optical-system design and in the processing of digital images?
6. (a) Create an expression to describe a simple phase object with a one-dimensional sinusoidal distribution. From this derive the features of the diffraction pattern from such an an object, and demonstrate how the diffraction may be modified (i.e. spatially filtered) to create an intensity object that is related to the original phase object, and sketch a suitable system.

   (b) There are at least two options on the approach to the above procedure – both mentioned in the text. Compare and contrast the features of these two approaches and also find examples of phase objects, some of which can be perceived with the naked eye. How is that possible?
7. (a) Convince yourself that the Bragg conditions (angle of incidence equals angle of reflection, and path difference between adjacent reflected rays

equals one optical wavelength; see Figure 4.12) can give 100% constructive interference for an input parallel beam meeting a suitable phase object. What are the conditions for the object to be ‘suitable’?

(b) A Bragg diffraction process uses ultrasonic waves in transparent media. The reflections involved are now from a moving object, so that the output beam is shifted in frequency according to the Doppler effect. Explain how this frequency shift is related to the ultrasonic frequency.

8. (a) Figure 4.14 shows the cross section of a dielectric waveguide comprising two low-refractive-index regions surrounding a slightly higher-index core region. Typical refractive indexes are in the region of 1.5 and typical index differences are around 1% of this. By considering the interference of intersecting beams at symmetrical angles with respect to the core axis in the waveguide, estimate the cross section of the guiding region at which the guide will become overmoded for a wavelength of 1 micron.

(b) The intersecting-beams approach works well for metallic interfaces in waveguides at much longer (for example, microwave) wavelengths but for dielectric guides, it can give only a good insight especially at these much higher frequencies. What are the differences between dielectrics and metals and how might these differences modify the original estimate for the depth of the waveguide?

5
Photonic Tools

In this chapter we look briefly at some of the many applications which are enabled through photonics, ranging from photo-therapies to intercontinental communications. The majority of these applications rely upon extracting an optical signal from various forms of background noise, which is where we start.

5.1 Optical Systems

When discussing an optical system the initial question is 'can we identify the parameter – the signal – in which we're interested'? Assuming it can be found, the next question is, 'how well can it be identified'? The answer to the first question is encapsulated in another question, 'does this parameter's presence exceed that of all the other phenomena which are happening at the same time'? Does the *signal* exceed the *noise*? The answer to the second question can be found in 'by how much does this signal exceed the noise'?

Noise in optical systems arises through a variety of mechanisms, depending upon the particular context. In an optical communication system, for example, an electronic receiver sees nothing but the modulated signal of interest. In contrast, imaging systems see the background illumination and spurious reflections which often compromise image quality. There is also the possibility that random noise originating within the light source will affect the signal of interest, and finally there are noise sources within the detection mechanisms themselves.

The simplest situation occurs when the light source of interest is incident upon a detector – typically a photodiode – in an otherwise darkened environment (Figure 5.1). Here there are inevitable sources of noise. First is the thermal noise within the load resistance on the photodetector. This rms noise power, P_{th}, is related to the absolute temperature, T, and bandwidth, B, through

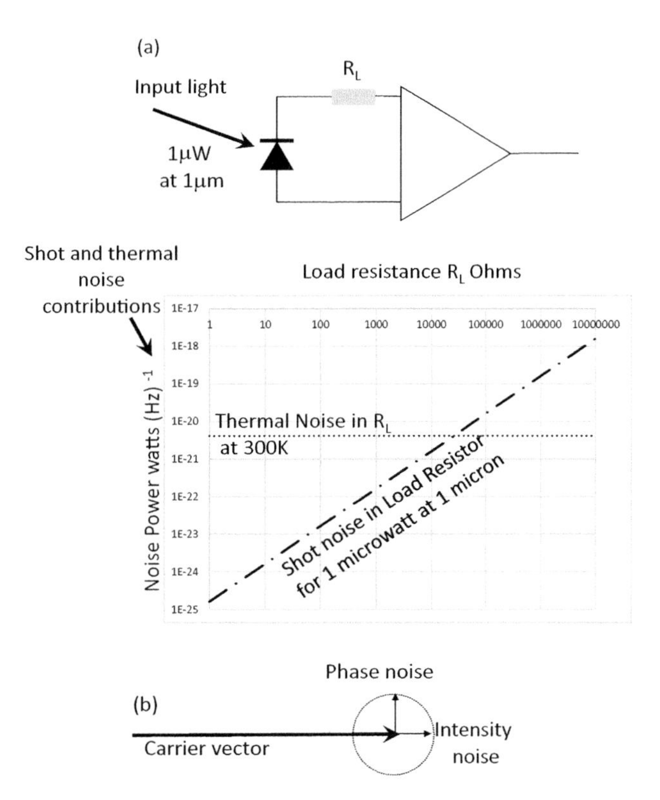

Figure 5.1 (a) A basic photodetector circuit demonstrating the trade-off between load resistor thermal noise and shot noise on the incoming optical signal. The load resistance and detector capacitance define the detection bandwidth. (b) A vector representation of noise components with equal distribution in phase and amplitude components.

the Boltzmann relationship $P_{th} = 4kTB$ (k is Boltzmann's constant). It gives a noise current dependent on the value of the load resistor.

The second noise source reflects the fact that the detector recognises the presence of the light as photons. These photons do not arrive at exactly regular intervals; they follow Poisson statistics, so that an average rate of arrival of N_p photons in a particular time interval will statistically fluctuate by $N_p^{1/2}$ when observed over successive time intervals. This is known as *shot noise*. Additionally, only a fraction of the arriving photons actually create electrons in the detector – a fraction quantified through a parameter known as the *quantum efficiency*. There will also be additional noise sources within the detection amplifier circuitry, not to mention inherent noise sources in the source of light itself, due to, for example, minute variations in the current driving the light

source. However, if we assume good design these effects can, in many practical cases, be considered negligible. Under such optimal circumstances and assuming 100% quantum efficiency, the ratio of the total current to the noise current within the photodetector is ${N_p}^{1/2}$.

In the ideal case – the shot noise limit – the noise due to the fluctuating photon arrival rate will dominate, so that, fundamentally, the noise performance of any optical system is determined by the average photon arrival rate over a time interval given through the reciprocal of the system's operational bandwidth.

This gives a straightforward and reliable indication of achievable signal to noise ratios. As an example, Figure 5.1 also includes a plot indicating the operational regime in which the shot noise on a 1 μW optical source at 1 μm wavelength will dominate system performance at the shot noise limit: this occurs when the induced shot noise current exceeds the thermal noise current in the specified load resistor. In the example, this requires a load resistor in excess of ~20,000 ohms.

This process can be applied to any photonic system which relies on electronic detection, but even then the impact of stray light can be a source of interference. Unscrambling the desired wanted signal from the unwanted signal in this more general case can be significantly more involved. There are also many occasions where the characteristics of the human visual system come into play, particularly in the context of displays and illumination. An emerging application sector in photonics showing immense promise lies in biomedicine.

Even in the shot-noise-limited case some subtleties need to be considered. The variability in photon arrival rate previously alluded to presents only a simplified perspective. Here, a complementary wave motion perspective, describing the same total noise power as a variation in the phase or amplitude of the mean wave vector (Figure 5.1(c)) is helpful. Thus the shot-noise behaviour can be viewed in the wave vector domain as an equally probable perturbation in the phase or the amplitude of the received optical signal, with equal power distributions in the phase and amplitude noise. Many optical systems rely on detecting changes in either the phase or the amplitude of the received optical vector; it can be helpful to retain the overall power signal to noise ratio dictated by the shot-noise limit but move it more towards the intensity sector for systems utilising phase modulation, or vice versa. An example follows.

The 'squeezing' of light uses elegant optics to 'push' the phase-noise component in Figure 5.1(c) into amplitude noise and hence to improve phase detection, or vice versa; this has been demonstrated using numerous approaches. This squeezing can in principle squeeze the phase, or the amplitude, noise away completely to zero and leave only the complementary domain. In practice, this has yet to happen completely. However, at present there is one outstanding example; it is being used in earnest on just one instrument, the gravitational-wave telescope, LIGO. Here, without optical squeezing giving a 10 dB or more additional signal to noise ratio, the detection of gravitational waves would have remained elusive. This gain is achieved through an impressively complex system, but to date there are no simpler approaches. Squeezed light exemplifies the many recent conceptual developments in photonics that combine the wave and photon approaches.

5.2 Spectrometers

Spectroscopy is a crucially important tool in photonics for characterising both materials and material structures. In practice, most spectrometers follow the format shown in Figure 5.2, where parallel white light is passed through the sample of interest to interact with the dispersive element (here a prism) after which the emerging light is focussed onto a detector array. The basic procedure is that the white light source is calibrated in the absence of the sample. Hence, any differences between the calibration readings on the array and those after the source interacts with the sample relate uniquely to the spectral properties of the sample. The dispersive element can in fact be either a prism or a diffraction grating; the former deflects the blue more than the red whilst the grating does

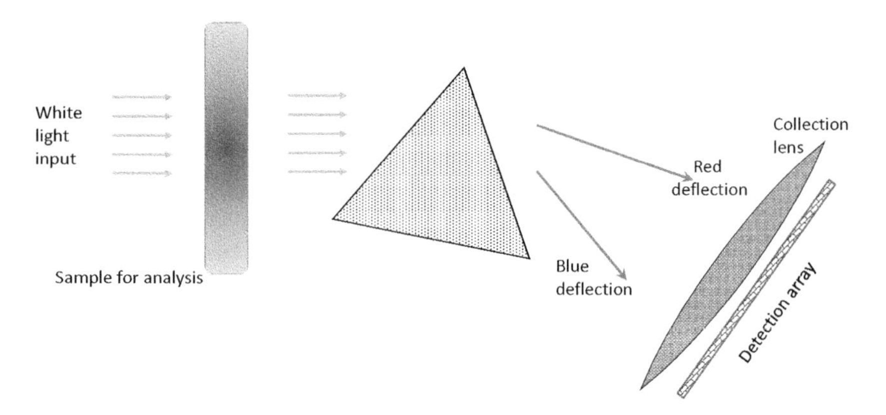

Figure 5.2 The basic dispersion diagram for a prism spectrometer. Grating spectrometers operate similarly but give the greater deflection to the red rather than the blue.

the reverse. A grating also offers the benefit that the angular dispersion is directly related to the grating period, whilst for a glass-based prism the relationships are more involved.

The intrinsic wavelength resolution of these instruments is dictated by several factors. The number of photodiodes in the detector array obviously limits the ultimate achievable resolution, (though with too many detectors, the achievable noise limit on each detector, which is shot noise limited, also contributes) Additionally, the angle of separation between one colour component and another comes into play. In the case of a grating this is determined by the grating periodicity whilst for a prism the dispersive power of the glass of the prism is the determining factor. Finally, the beam width of the interrogating optics (comprising a combination of the beam width, the depth of the sample container, and the effective incident width) provides an ultimate limit, dictated by diffraction from the edges of its optical aperture. In an optimal design all three of these limitations would come into play together.

There is also an important variation on the basic theme of the photodetector array. The same functionality can be achieved if a single photodiode, with a slit in front of it, is scanned across the resulting spectrum either by moving the photodiode or by rotating the spectrum. The overall result is that, for a typical prism or diffraction grating spectrometer, a spectral resolution of the order of 0.1 nm over a dynamic range of 1 μm is achievable.

There are occasions when better resolution is desirable and also occasions where the photodetection processes become more involved; typically, both factors apply in the near and mid infrared, where an alternative approach based on Fourier transform spectroscopy (Figure 5.3) is frequently used. In this system a single detector records the total output as the path difference between the two arms of the interferometer formed by the two mirrors is scanned from around zero to a maximum value. At each setting the interferometer acts as a filter with co-sinusoidal response, as indicated in the figure. Consequently, in effect each optical frequency in the input spectrum is multiplied by this co-sinusoidal function in the frequency domain for each setting of the interferometer path difference – in other words, the system is performing a Fourier transform at each setting with spectral components determined by the path difference. Assuming that the path difference increments by equal amounts between a minimum (which sets the range of frequencies that can be resolved by the interferometer) to a maximum (which determines the resolution of the resulting spectrum), then a direct Fourier transform of the output gives the spectral response of the sample. The detector needs to be calibrated appropriately in the range over which it is to used, and some precautions are needed to ensure that the Fourier transforms are appropriately performed, taking into

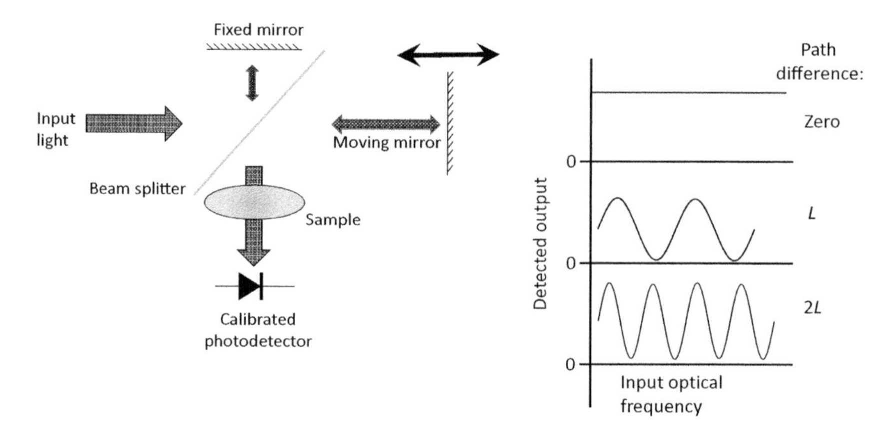

Figure 5.3 Fourier transform spectroscopy, showing how varying the position of the moving mirror by L multiplies the input spectrum by a defined co-sinusoidal function. The beam splitter divides the input light equally between the two paths through the interferometer. The top line on the graph indicates that at zero path difference the output power is fixed for all wavelengths.

account any spectral variations i.e., variations in the light output from the interrogating optics as a function of wavelength. Overall then, the system is functionally identical to the dispersive element in a prism or grating spectrometer.

5.3 Display Technologies

Display technologies manifest themselves in two basic formats. The first is the familiar luminescent screen, to be seen on TVs, phones, tablet, computers and the like. The second uses an external light source to illuminate a variable, usually transmitting, object; such spatial light modulators are most familiar in the operation of digital projectors. The details of these technologies are many and varied, though essentially luminescent displays rely upon light emitting diodes, whilst spatial light modulators rely on liquid crystals and usually incorporate appropriate colour filters.

Naturally, display technology needs to be designed with the human visual system in mind, usually targeting the daylight-adapted eye, which resolves colour through cone receptors. These receptors are sensitive over a wide range but individually peak in the red, the green and the blue, giving rise to the concept of primary colours from which all other perceived colour combinations can be synthesised. As a group, the receptors have the overall spectral response typified in Figure 3.19. This synthesis process from the three primary colours was quantified nearly a century ago in the CIE (International Commission on Illumination) chromaticity chart, still a very useful tool in the detailed design of visual systems.

These factors are mentioned here as among the most important considerations in the design and implementation of both display technologies and also electronic image acquisition technologies, such as the image capture array in a digital camera. Hopefully this will give the reader a little insight into the considerable efforts required in balancing colour-filtering systems, relative brightness, spatial resolution and a host of other factors in the design of these very commonplace but extremely subtle devices.

5.4 Images and Imaging

All image formation, whether by a telescope, a microscope or even a magnifying glass, is based upon the same principle. A lens collects light from the object and thereafter projects an image which is subsequently processed by the viewer and received via the retina before being processed by the brain. There are often many steps in the process – camera to data file, to image on a screen, to the viewer, for example. Images, as we have already discussed qualitatively, can be decomposed into Fourier components, a representation of which is found in the back focal plane of the collection lens. This Fourier transform plane is widely exploited as a means for image processing using both direct-optical and digital techniques. As examples, we have already mentioned the concepts of contrast enhancement and phase-object-to-intensity-image manipulation.

5.5 Three-Dimensional Imaging

The basic principles of holography were outlined in Chapter 4. It emerged in the mid twentieth century as a concept which was facilitated through the evolution of the laser. The basic observation which underlines holography is that if we can somehow record both the phase and the amplitude of a propagating optical wave, we can determine whence it came. Photography, whether digital or on film, like our eyesight perceives only the intensity of the arriving wave as a function of position, producing the familiar two-dimensional images. The key point, then, is to change phase information into some form of intensity information which can be recorded, and thereafter to find a playback mechanism.

Holographic systems have evolved from their original novelty as a means of producing strange and intriguing three-dimensional images which just hang in space (numerous science museums continue to excite young minds through this very artefact) into useful tools, derived from variations on the basic principle, for surface inspection, vibration detection and similar uses. The principle of this use of holography is as follows: record a hologram of an object which is to form the reference and place the sample object exactly in the position where the reference object is reconstructed. Any differences between the reference object and the sample object will then appear in the form of interference fringes, the detectable differences being of the order of a fraction of a wavelength.

Holograms make their daily appearance as security checks on credit cards. They are being evaluated for extremely high-density data storage – especially in the more versatile volume-hologram variant, where the propagating wave front phase distribution is recorded not in a two-dimensional plane but through the thickness of an emulsion extending over many wavelengths of light.

In common with many photonic concepts, holography has evolved from an interesting scientific principle into a widely applied tool, ranging from entertainment, to high-precision surface inspection, to security and identification checks.

Optical-coherence tomography (OCT) presents an entirely different approach to deriving a three-dimensional image. The principle is shown in Figure 5.4. The object is illuminated using a very low coherence length source – typically a super-luminescent diode. A reference mirror can be moved along the optical axis as indicated and the consequent interference pattern seen by the receiver will produce peaks and troughs only when the optical path difference between the imaging arm and the reference arm is around zero, thanks to the very short coherence length of the super-luminescent diode. Consequently, recording the image as a function of this path difference gives the reflectivity of two-dimensional slices of the object as a function of depth. Stacking together these slices gives a three-dimensional image of the object. The depth for which this system can operate is effectively dictated by the range of path differences which the moving mirror can access and is limited by attenuation and scattering within the object under investigation.

The system of optical-coherence tomography has found extensive application in producing diagnostic images of the eye, and the apparatus is found in many high street opticians. It has also been used intravenously for monitoring cardiovascular degeneration and, in an entirely different domain, as a tool for imaging and assessing silicon and silica device-fabrication processes. Thus OCT has grown to be an important and versatile three-dimensional imaging tool since the original concepts were identified in the early 1980s.

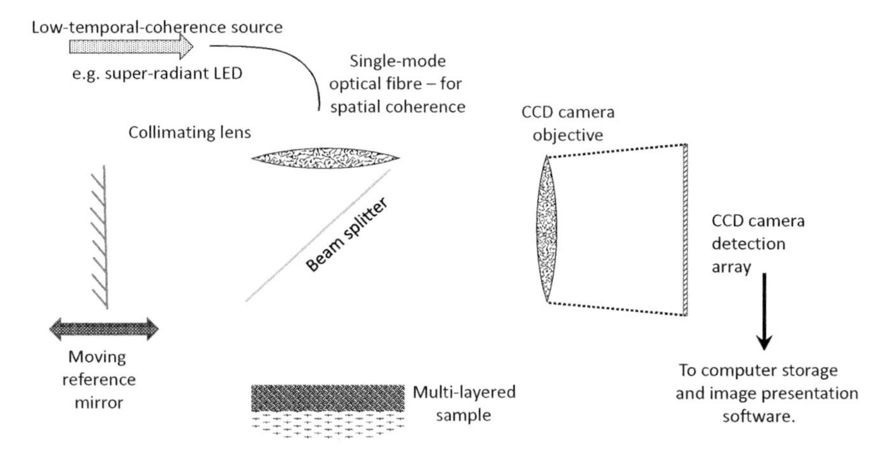

Figure 5.4 The basic principles of optical-coherence tomography. The moving reference mirror selects a plane within the object which will produce constructive interference and give an image on the camera array corresponding to the structure within that plane.

Predating these techniques were the – still used – three-dimensional imaging systems exploiting the fact that we can see in three dimensions owing to subtle processing of the differences between the images seen by each of our two eyes. If we take two simultaneous photographs using cameras spaced by the same amount as a person's two eyes then, using a projection system which sends one image to each eye, the two images will be perceived as a three-dimensional image. The early three-dimensional cinema operator projected orthogonally polarised images for each eye and equipped the audience with left and right eye selective glasses tuned to the two polarisation states. Three-dimensional virtual reality systems are based on the same dual-image photography concept.

5.6 Precision Measurement

Optical interferometry, through which movements of the order of a small fraction of an optical wavelength can be reliably detected, is a natural tool for precision measurement. The application sectors have gradually been expanding for over half a century.

As we have seen, interferometers measure changes in the optical path difference between two points. In the case of a two-beam structure, this path difference is between a fixed reference object and some other moving reflective structure. For the simplest two-beam interferometer, in which the coherence

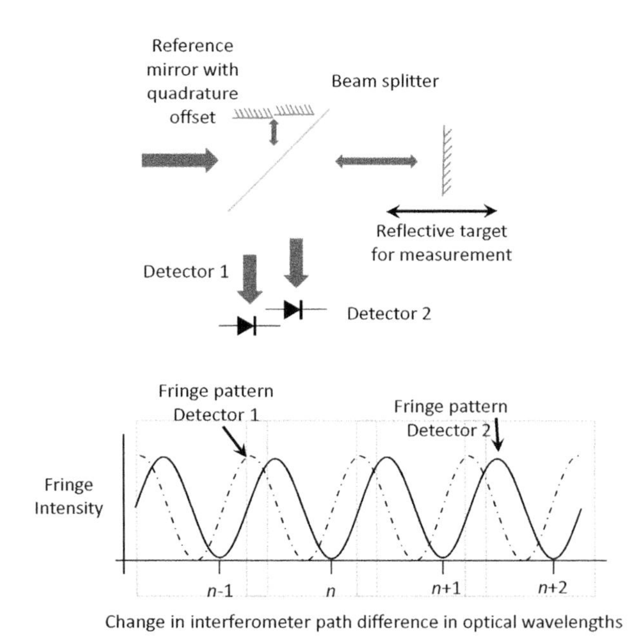

Figure 5.5 Using a quadrature offset system (i.e. the beams are 90° out of phase) in a Michelson interferometer to extract directional information and therefore measure absolute changes in position.

length of the source greatly exceeds any possible value of the path difference between the two arms, one cannot distinguish whether the path difference is increasing or decreasing as it monotonically changes. However, a change in the direction of movement can be detected from a single out put provided that this change does not occur at a peak or trough of the interference pattern (see Figure 5.5). Consequently, interferometric precision-measurement systems need to incorporate some method of determining the direction of changes in path difference. One approach, shown in the figure, is to utilise a quadrature offset between, in effect, two reference beams. This arrangement enables a combination of fringe counting and directional assessment to produce highly accurate measurements. This approach, based upon quadrature detection, can routinely yield measurement precision better than 1% of an optical wavelength; measurement accuracies can even be of the order of several parts per million, depending upon the stability of the source illumination and the variations in local environmental conditions.

Interferometry has found enormous application in the precision measurement of mechanical objects, especially in systems based upon coordinate-measuring machines. In these machines, a very precise *x*–*y*–*z* translation system is coupled to interferometric displacement-measurement systems. The translation system

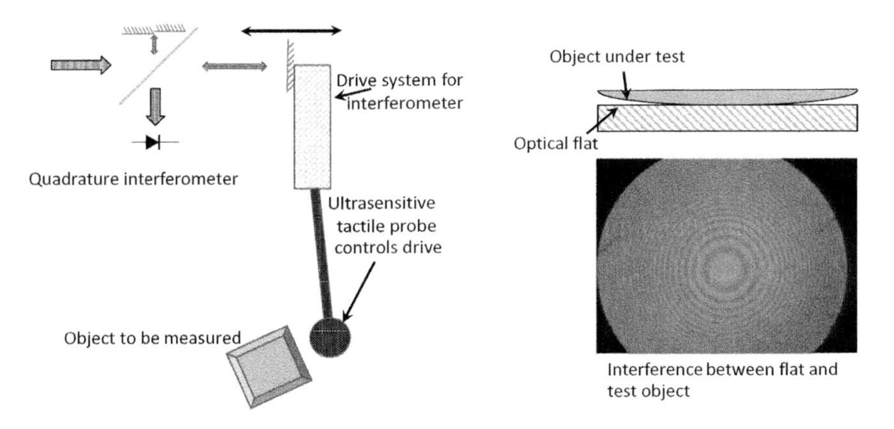

Figure 5.6 (left) Measuring a fixed external object using a sensitive probe and quadrature interferometer. (right) Using interferometry to measure objects with respect to an optically flat surface (here interference produces Newton's rings, which are observed from a spherical partially reflecting surface as indicated). The same basic approach is used in holographic inspection systems.

carries a very sensitive tactile probe which responds precisely as it touches the object under study. There are several such probes, at least three in a particular instrument, to facilitate three-dimensional measurements. The displacement of these tactile probes can be interferometrically monitored and from this the surface profile of the object can be precisely determined. Submicrometre resolution can be routinely obtained.

Figure 5.6 illustrates this basic tactile system and also shows how a precision optically flat glass can be used to characterise the shape of a lens. Now referred to as Newton's rings, this patteon was first observed over 300 years ago. When viewed with white light the rings are coloured, but when viewed with monochromatic light, for example from a gas discharge lamp or even through the rudimentary colour filtering of sunlight, the characteristic interference pattern shown in the figure provides an accurate measure of the lens curvature, and variations thereof, with respect to the reference optical flat. The spacing of the fringes depends upon the local tangential angle between the lens surface and the optical flat.

Yet another variation on the interferometric measurement theme is shown in Figure 5.7. The Sagnac interferometer sends light in counterpropagating directions around a single-spatial-mode loop, which can be a fibre or the active part of a laser. As the loop rotates, the light travelling in the same direction as the rotation needs to travel a small distance further than the light travelling against the rotation in order to return to the entry point. Consequently, if this system

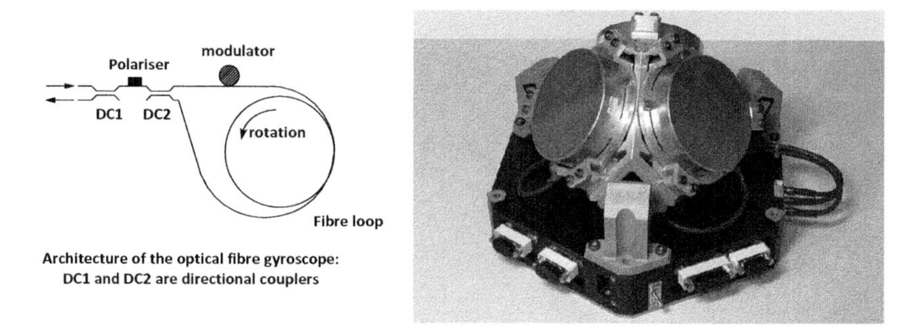

Figure 5.7 The basic principles of the Sagnac interferometer, as used as an optical fibre gyroscope, and an example of its precision realisation for space navigation.

forms the cavity of a laser, there is a small difference in oscillation frequency for each direction of travel. In the fibre-optic case shown, there is a corresponding small phase difference. With subtle and elegant signal processing, this phase difference can be measured with a precision of a few nanoradians, which translates to a rotation rate sensitivity of less than 10^{-4} degrees per hour. This is sufficient to guide space exploration and two orders of magnitude better than that required to navigate an aircraft across the Pacific Ocean. In the laser case, there are issues in separating the clockwise and counterclockwise lasing systems when the resonant frequency differences become very small (the two separate oscillators become coupled to each other and lock onto a single frequency), so that this variant on the Sagnac gyroscope tends to have a limit in the region of 10^{-2} degrees per hour.

These brief examples merely scrape the surface of the uses of interferometry in measurement but are intended to give some indication of the precision with which interferometric measurements can be made.

5.7 Optical Fibres and Communication Systems

The basic concepts of optical fibres were first mooted more than half a century ago. The initial attraction was that losses below 20 dB/km appeared feasible provided that the glass, strictly silica, was extremely pure and the interface between the core and the cladding was sufficiently smooth to prevent scatter.[1] Copper cables for electronic systems in communications were then reaching

[1] The term 'scatter', abbreviating 'the scattering of electromagnetic radiation' is often used when we are considering a phenomenon rather a designed process.

their limit at around this attenuation level for carrier rates in the 250 Mbits/s region. However, the skin effect losses (see Figure 3.3) in copper increase with frequency, yet even higher frequencies were needed to carry more and more data. As we saw in Chapter 4, numerous paths in the simplest fibre will continue to guide until cut off at the critical angle between the core and the cladding. This range of paths implies a travel time difference which is typically of the order of a few tens of nanoseconds per kilometre. This dispersion in turn implies the need for maximum modulation frequencies into the fibre in the region of many tens of megahertz on a one-kilometre length. Thus, initially multimode fibres offered little obvious benefit for long-distance communication when compared with copper.

However, there is a modification which changes the picture somewhat. Suppose the refractive index profile were to change more gradually than the abrupt high-in-the-core and low-in-the-cladding version profile found into a step-index fiber. If the profile were instead *parabolic* (along the same principle as the parabolic mirror in an astronomical telescope) then, at least for one specific wavelength, the dispersion effect could be removed. There are practical issues regarding how closely can we make the refractive index profile into a parabola, and there is also material dispersion in the core itself, but the graded-index multimode fibre remains a useful tool. You may wish to explore further how this might work and why it could be useful.

Single-mode fibres offer improvements in dispersion over even a graded-index fibre. However, even here the minimum-attenuation wavelength (1.5 microns) and the zero-material-dispersion point at 1.3 microns do not coincide. Making the zero-dispersion point coincide with the very low attenuation region then promises enormous benefits!

The 1.5-micron lowest-attenuation region has emerged as the optimum wavelength for optical communications, and so minimising dispersion is critical. Considerable ingenuity has been directed at this problem, based on the observation that the evanescent tail (see Section 4.2) of the optical field distribution in the fibre cladding extends further into the lower-index cladding as the wavelength increases. Consequently, that mode will have a faster velocity as it penetrates further into the lower-index region. A carefully designed 'W' refractive index profile (Figure 5.8) controls this wavelength-dependent evanescent field by means of a high-index core followed by a lower-index cladding surrounded by a slightly higher-index cladding. This design can in fact compensate for the material dispersion; it controls the relative amounts of optical power in the core and depressed cladding regions and the process is optimised to realise this objective. The reader can consult Figure 4.16 to verify the logic behind this approach. The result is extremely low-loss transmission (a small fraction of a

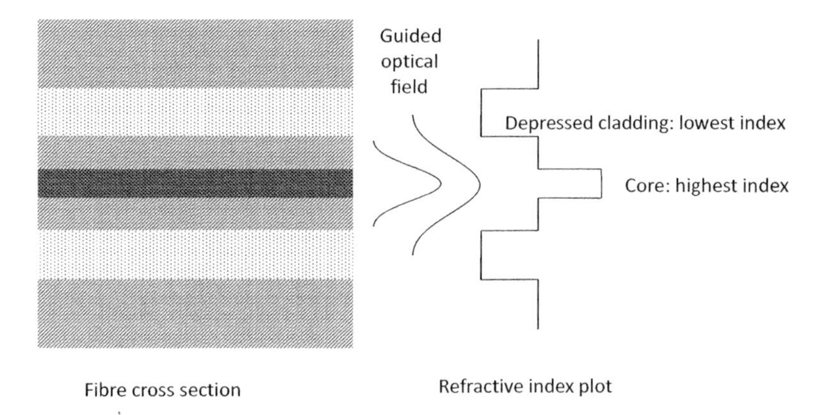

Figure 5.8 The basic concept of dispersion compensation, combining the prospect of low attenuation at 1.5 μm whilst correcting for finite material dispersion (see also Figures 4.14 and 4.15). In the central part of the figure are shown a shorter-wavelength pulse and a longer wavelength pulse, which penetrates further into the depressed cladding region.

decibel per kilometre) and the residual dispersion is measured in picoseconds per kilometre. This performance is critical in enabling the unprecedented capabilities offered through the current communication systems and the enormous internet bandwidth capability now taken for granted throughout the developed (and increasingly in the developing) world.

5.8 Lasers in Machining and Surgery

The inherent spatial coherence of a laser beam has many benefits, of which possibly the most evident, and definitely among the most remarkable, is that straightforward focussing optics can produce spot sizes with dimensions in the microns. Huge power densities are then easily attained; a one watt laser beam focused to a 10 μm diameter produces in excess of 1 MW/cm^2 power density. Lasers with power in the hundreds or even thousands of watts are now available, upping the available focussed power densities to GW/cm^2.

In almost any material this will produce huge, highly localised temperature rises, of the order of thousands of kelvins, with melting or even evaporation in the very small focussed-beam area. Even at much lower power densities, significant changes in highly localised structural (bio)chemistry can take place.

The applications of these small focused spots are diverse and Figure 5.9 indicates some of the prospects. The most obvious is laser machining, both cutting and welding. Such laser cutting enables consistent precision; there is no

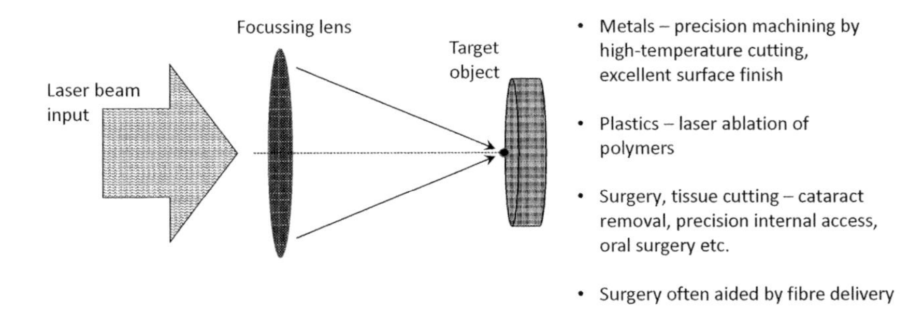

Figure 5.9 The essentials of laser machining and laser surgery, utilising a tightly focussed spot from a high-peak-power pulsed laser, very similar to laser-induced-breakdown spectroscopy (LIBS) but with a different aim.

cutter wear and minimal machine vibration. The process also offers smooth edges and the ability to cut in otherwise inaccessible places, particularly with the flexibility of fibre-coupled cutting systems.

Laser surgery has had a particular impact in the treatment of ophthalmic problems such as macular degeneration and cataracts where highly localised cutting and preparing without the use of the surgeon's knife is particularly beneficial. The laser scalpel has also found its place, especially in soft-tissue surgery where its precision and the minimal consequential bleeding that it produces, owing to self-cauterisation, offer great benefits. Also, a laser beam clearly does not require careful sterilisation between procedures, though there is the need to pay due attention to the delivery system, which could even be a disposable fibre. Laser surgery has been incorporated into endoscopically observed surgery and a host of other medical procedures.

Lasers have also found their way into dentistry. There is a clear optical distinction between decay in a tooth and the intact regions around it, with the former much more optically absorptive. There is then the potential for thermal removal rather removal by drilling. Lasers are also useful in oral surgery, again especially when fibre-coupled to facilitate the precision location of a cutting tool, and this, together with the above-mentioned automatic self-cauterisation, are attractive features. Indeed, possibly the only disadvantage of lasers is that the capital outlay compared with a scalpel or a mechanical drill remains high.

5.9 Environmental Analysis

This embraces one of the most important application sectors for spectroscopy in its many forms. Measurements using absorption spectroscopy, Raman or

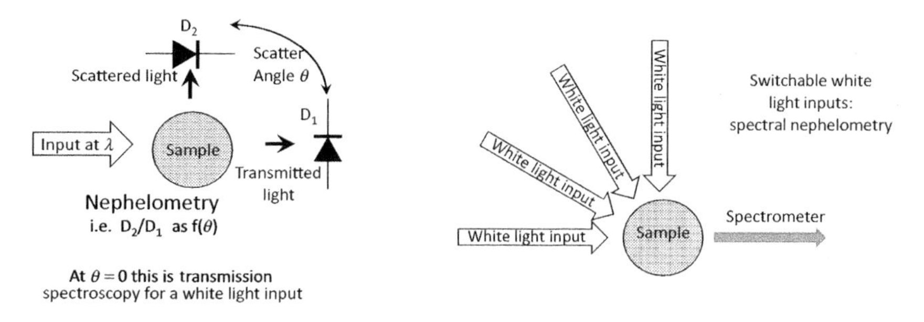

Figure 5.10 Spectral nephelometry characterises samples (usually liquids) for both absorption and scatter as a function of angle. The detector diodes are indicated by D_1 and D_2.

laser-induced-breakdown spectroscopy (LIBS) reveal much about the air that we breathe. Particulates are present in air, and light scatter reveals much about this form of atmospheric pollution. Sometimes it is obvious in the smog which affects many cities worldwide. The atmosphere is, however, the simplest puzzle to solve optically. Absorption spectroscopy can uniquely identify nitrous and sulphur oxides, for example, and can accurately measure the levels of carbon dioxide. Particulate scatter is also straightforward to measure and readily characterised.

Solids and liquids present different issues in that – as we have already discovered – the spectra of individual elements within a particular compound, and those of the compositional components in a mixture, overlap, so that direct spectroscopic measurements are much more difficult to interpret. Figure 5.10 shows one of the many ways around this difficulty, which is based upon measuring a combination of scattering signature and absorption spectrum coupled into elegant signal processing – a technique known as *spectral nephelometry*. This has enabled differentiation among olive oil samples, beer, whisky and wine and is becoming accepted as an evaluation tool (Figure 5.11).

Water analysis presents other challenges, many associated with its almost total transparency in the visible. There are, however, important features in both the near infrared and the ultraviolet. Additionally, scattering can give an indication of particulate contamination, though bacteriological characterisation requires, thus far, the biochemical analysis of samples. Nevertheless the future may well hold other prospects, involving for example induced fluorescence, optically active test bacteria seeking complementary species or advanced imaging.

Hyperspectral imaging is another important technique, particularly for land and atmosphere characterisation. The technique entails taking the spectral

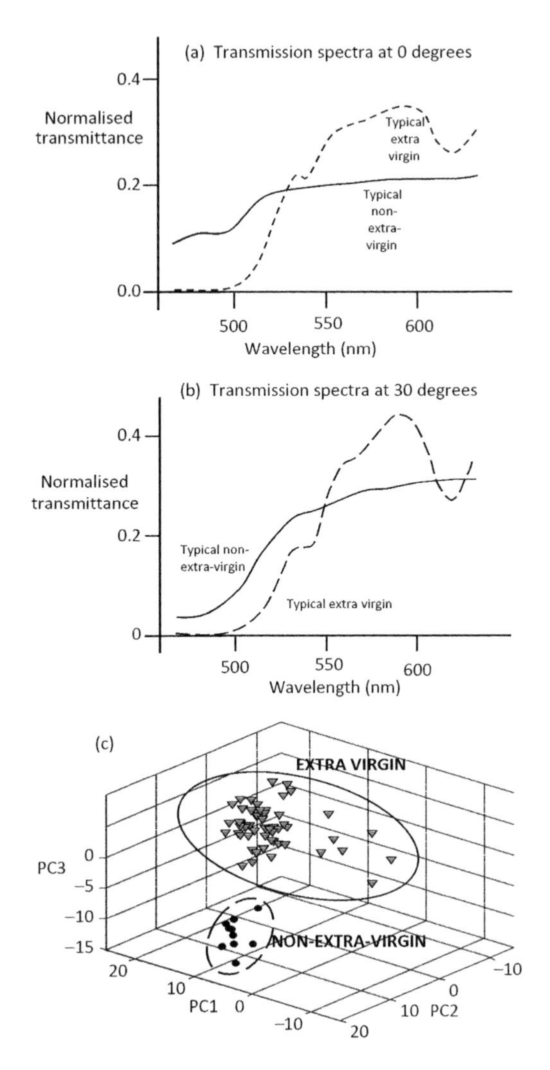

Figure 5.11 Extracting information from nephelometry data using principal components analysis to separate extra virgin and non-extra-virgin olive oil. There are many applications for the technique, and numerous signature identification algorithms.

signature, in the limit, pixel by pixel, of a photographic image. As an example, recall that many gases have significant absorption spectra in the near infrared. Methane absorbs strongly at about 3.3 μm wavelength – a wavelength at which there is considerable solar illumination. Suppose, then, we take a picture along a gas pipe with a special camera (FLIR, forward looking infrared) which is filtered to look at this specific wavelength only. If there is an actual gas escape then this will absorb the background radiation at the position of the leak and appear as a black cloud. This idea is used in practice, and not only for methane but for many other species. The principle can also be generalised to examine, for example, crop distribution through satellite imaging by chromatically characterising a reflected spectrum to give a plot of land fertility, arid and moist regions, or the distribution of crops. Environmental assessment currently features among the principal applications of photonic technologies.

5.10 Light as a Therapy

Biomedical photonics has become an increasingly important discipline and continues to expand as the many biological and physiological influences which light can have become more accurately characterised. Many diseases of the skin respond constructively to ultraviolet light, typically in the 250–400 nm band. There is also evidence that appropriate photonic treatment can enhance the healing of cuts and bruises and assist in skin cancer therapies. More recently neurological possibilities have been explored, where very precisely positioned optical probes, typically fed through fibres and coupled into the appropriate source wavelength, can implement beneficial neurological changes in the brain.

The influence of light and spectral distribution on sleep patterns, not least as affected by jet lag, and on sleeping disorders, is well documented, the red and orange part of the spectrum (with the lowest frequencies) being much more restful than the higher-frequency blues and violets. (Perhaps the reds and oranges are reminiscent of a soothing sunset?) Ultraviolet light has been used as a sterilising agent both in water treatment and in bacteriological control in humans. This just gives an indication of what is currently happening in phototherapy, which promises to become an immensely productive area for future research and application.

5.11 Astronomy and Very Large Telescopes

Stars have been a source of fascination since ancient times, driving a desire for ever improving tools to observe the distant universe. The essential issue remains how to get the maximum light from these distant worlds to observe details beyond the capability of the naked eye whilst simultaneously minimising the ‘twinkle’ which results from even very minor atmospheric perturbations. Gathering the maximum light implies greatly expanding the collection area, whilst maintaining highly precise aberration-free optics, which in turn inevitably increases a telescope's sensitivity to atmospherically induced twinkles. The requirement for large aberration-free optics leads to the use of mirrors rather than lenses, which both simplifies the surface preparation and minimises the overall system weight and thereby improves a system's rigidity. Minimising the twinkle points towards taking apparatus up mountains or indeed into space.

Mount Palomar in California was the first of these high-attitude locations, at about 1700 m altitude and with a 5 m telescope. Both the height and diameter of such telescopes have increased since then, and Mauna Kea in Hawaii at 4200 m hosts a range of instruments with mirror diameters to 10 m. The principal secret of all these telescopes, and indeed of the astronomical equivalents such as the Hubble and the Webb, lies in achieving the necessary precision over a huge area in order to produce a parabolic surface for large-aperture high-resolution imaging; the mirror should not deviate by more than a fraction of a wavelength from an ideal parabola over its entire surface. Simultaneously, the necessary supporting framework, compatible with transportation along a difficult route, whether up a mountain or in a rocket, must be realised. The mechanical design must both withstand rough handling and also retain its shape through all the operational conditions to which the telescope will be subjected. Whilst the basic optics shown in Figure 4.6 illustrates the principles, it belies the huge feats of mechanics and precision assembly necessary for a successful implementation.

There have been enormous improvements in the basic design (Figure 4.6) of such telescopes over the past few decades, of which perhaps the most important aspect concerns the adaptive optics (Figure 5.12), which facilitates very precise mirror shape control to correct for thermally induced strains and other mechanical stresses, as might occur during the launch of systems designed to be extraterrestrial.

These adaptive optical systems also offer another important benefit in addition to fine tuning the initial mirror machining: they are used to correct for atmospheric perturbations and to ensure that the telescope remains focussed on a particular star over long periods, to give the longest possible exposure in any photographic record. The essential concept behind both these functions is the 'guide star': the telescope motion system and the adaptive optics are both guided by a particular celestial object. This guide star needs to be bright enough to be detected rapidly and hence to provide input to the feedback system, and it also needs to be close to the area of interest on the sky. In the absence of a bright guide star, a terrestrially located laser beam provides

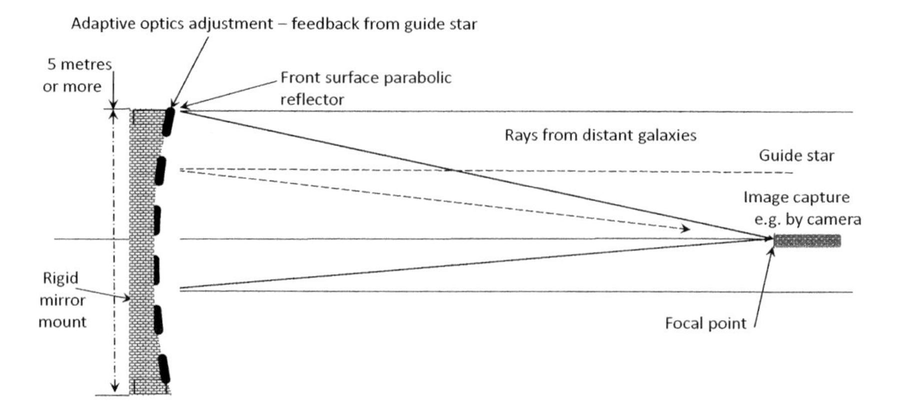

Figure 5.12 Adaptive optics precisely adjusting the parabolic reflector in, for example, a very large space-borne optical telescope.

similar functionality, especially in correcting for atmospheric aberrations. Tuning the pointing direction of the telescope then becomes a little more involved, but combinations of celestial patterns, a less bright guide star and precision navigational instruments to lock onto the reference-laser guiding beam can ensure that everything functions well.

Nevertheless, the astronomical community continually yearns for more light gathering, that is, for even larger telescope areas with the consequent enhancement in image resolution and the corresponding need for even more stability in image acquisition. Venturing into the extraterrestrial is an inevitable consequence, not only in the search for ever more detail in the astronomical map but also in endeavouring to answer questions about extraterrestrial life. The James Webb Telescope coordinated through NASA is an example of such a project in progress. The Webb will have a 6.5 m primary mirror and will orbit a million miles from earth, with all that implies – especially from the presence of solar heat and cosmic rays. There are also terrestrial aspirations – the European Extremely Large Telescope, with 39 m primary mirror, is finding a home high in the mountains in Chile.

5.12 Optical Tweezers

Optical tweezers have evolved since they were initially demonstrated towards the end of the last century into a much used tool in biomedical sciences and similar applications involving the manipulation and spatial stabilisation of tiny particles such as biological cells. The essential idea involves the carefully controlled

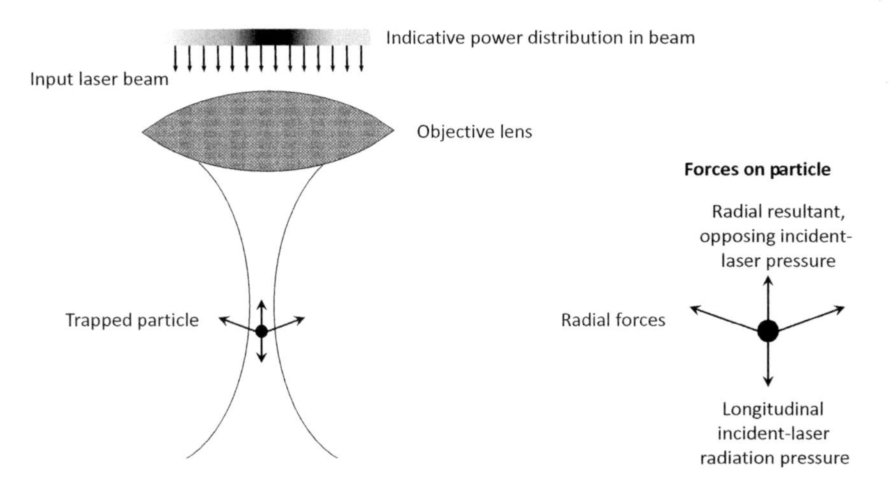

Figure 5.13 The basic principles of optical tweezers. Just below the waist of the beam the longitudinal radiation pressure on a particle is balanced by the longitudinal opposing component of the radial forces due to refraction through the particle; the momentum transfer is against the photon flow and can balance it.

utilisation of radiation pressure and the gradients thereof in a focussed laser beam, as indicated in Figure 5.13. Radiation pressure occurs when a photon, of energy $h\nu$ and momentum $h\nu/c$, is reflected from a surface of a particle; then the particle feels the rebound as a force $2h\nu/c$, corresponding to a pressure $2I/c$ where I is the incident light intensity (the optical power per beam area).

The forces on the particle from the radiation pressure will increase towards the focal point of the lens and there will be a cross sectional gradient tending to keep the particle in the centre of the beam. As the light strikes the particle, some will be refracted through it and will emerge at a different angle, imparting a momentum change to the particle equal and opposite to the momentum change of the light. When the particle is just below the focus (or waist) of the beam, its momentum change has a component *towards* the laser source and under appropriate conditions the radiation pressure and refraction forces balance to suspend the particle in the beam (for further information see the Wikipedia article on optical tweezers).

The forces involved are very small, in the 100 nano-newton range, and the focussed spot is typically produced by a high-aperture microscope objective lens with input laser power in the mW range. Particles with sizes from a few microns down to 5 nm can be trapped, facilitating the examination of objects ranging from biological cells to DNA molecules.

Optical tweezers have evolved into many formats since their inception over 30 years ago. They are used in microfluidic experimentation to trap and examine cellular structures in biological assays, to furnish microfluidic pumping systems, for trapping microscopic liquid droplets, for trapping and examining particles in Brownian motion and for a host of other manipulative functions on the microscopic, or even nanoscopic, scale. Radiation pressure is a very versatile tool.

5.13 Summary

This applications chapter has indicated the diversity and versatility of photonics as a tool, as an enabler, as a corrective treatment. Of necessity, much has been omitted and all the discussions had to be brief. Nevertheless, hints have been given of the roots of the large expansion in photonics since the mid twentieth century, The discipline is merging into everyday life, with DVD players, fibre optics to the home, powerful photographic tools and displays, the laser pointer, laser surgery and machining, photonic therapies and photonic systems in the local opticians.

There is also is a great deal more emerging for the future. The functionalities offered by photonics will expand enormously. The fabrication tool set, especially for extremely high-precision machining at the nanometre level, and also the conceptual framework for future prospects, will broaden significantly. We shall briefly explore some of these emerging possibilities in the next chapter.

5.14 Problems

1. The bandwidth of an optical receiver is basically determined by the product of the photodetector's capacitance C and its load resistance R. A bandwidth of 1 GHz is well below what is achieved in many optical systems.

   As an example, take beam with received power, at 1.5 micron wavelength, of 1 microwatt with 100% quantum efficiency. Now, arrive at some reasonable values for R and C for this bandwidth remembering that the thermal noise current in the resistor needs to be below the shot noise current for optimum performance. Look up achievable values for the junction capacitance and comment on (a) whether that would be all the relevant and unavoidable capacitance at 1 GHz, and (b) any trade-off between the necessary load resistances, the desired bandwidths and the received shot-noise-limited optical power, after arriving at a reasonable operational signal to noise ratio for the detection system.

2. Grating spectrometers are regularly used for many types of spectral analysis. You are seeking a spectrometer with an operating wavelength range from 1 micron down to 200 nm with a resolution of 0.5 nm.
   (a) What would be needed in terms of the detection array, i.e. the number of sensors in the charged-couple device (CCD) detector, the aperture of the input lens and the power spectral density of the illuminating white light source? What would be an acceptable signal to noise ratio for each detection point? Remember that you will need to quantify the attenuation in an absorption band with an accuracy of, say, 1% of the input source in order to achieve this performance level.
   (b) How might the signal to noise ratio be enhanced without changing any system elements (except in the detection process – assuming that time is not too pressing), and also what might be the implications of taking the desired resolution down to 1 pm?
3. Fourier transform infrared spectroscopy (Figure 5.3) is noted for its potentially high resolution. Consider a spectrometer that is required to operate in the range 0.5 to 1.5 micrometres with a resolution of 10 pm over that range.
   (a) What would be the necessary travel distance necessary for the moving mirror? From your answer make some estimates – with reasons – of the ultimate resolution of such a system.
   (b) Could it go to GHz resolution, or better, for example?
4. Optical coherence tomography (OCT) (Figure 5.4) is a widely used tool.
   (a) For an optical source operating at 1 micron what would be the minimum spectral width in nanometers needed to achieve a depth resolution of 5 microns? Would such a source be practically available?
   (b) From your observations comment on the practical limits in depth resolution which may be achievable. Assuming that the necessary precise (how precise?) mirror drives are available, what could restrict the possible depth range through which an image may be constructed?
   (c) The major applications of OCT lie in medicine, predominantly ophthalmology, but OCT has been used in, for example, measuring the thickness of a silicon wafer. Discuss – with reasons – whether the same system which works for silicon could also be effective for measurements on the eye.
5. The fibre optic gyroscope (Figure 5.7) relies on the fact that light emerging from the fibre travelling in the direction of rotation will stay in the fibre for slightly longer than the light travelling in the other direction. It is based on the Sagnac effect, originally described over a century ago.
   (a) Using this basic observation, arrive at a simple expression for the phase change in a fibre gyroscope as a function of the source wavelength, the

fibre length, the coil radius and the rotation rate. Then check to see whether your answer is appropriate (there are good web sources for this).

(b) From this expression make some observations on what is needed to obtain a sensitivity of 0.0005 degrees per hour in a loop the size of that indicated in the figure (you can work out the rough magnitude of the dimensions from the cables and connectors).

6. This problem concerns the underlying principles of optical tweezers. Suppose we have a laser beam of total power 1 mW focussed through a microscope objective to a spot 2 microns in diameter.
   (a) Make an estimate of the radiation pressure force exerted on a disc one micron in diameter placed in the centre of the beam.
   (b) From the above estimate, indicate how these forces might distribute themselves around a spherical object of diameter 1 micron.

These estimates will give an order-of-magnitude insight into tweezer operation.

6
The Future

Light, as we have seen, has being intriguing features. We think of it as generated and detected as particles, namely photons, but these particles are transmitted as waves (or sometimes can be accurately modelled as rays). The wavelengths of these waves are typically submicrometre. Most of the current prospects for photonic systems lie in combining the concepts of electromagnetic waves and photons with the ever expanding capability to manipulate materials and material structures to optically subwavelength tolerances. Specifying dimensions to nanometre precision is no longer wishful thinking.

The electromagnetic properties of materials, and the ability to use these properties in conjunction with structures at subwavelength tolerances, have already had a profound effect on the evolution of 'conventional' electronic systems. The ever increasing complexity of integrated circuits, the evolution of the compact mobile telephone and a host of other, now commonplace, devices have all been enabled by precision structural machining. A piece of copper wire has varying electromagnetic properties depending upon the frequency of the current going through it and its geometry; the simplest example of this is the contrast between a coil and a straight wire. A vital difference in photonics, however, is that whilst at lower frequencies generation and detection are best viewed as involving electric currents, for photonic frequencies the generation and detection of the electromagnetic wave as a photon must also be taken into account. This opens up a whole new dimension of possibilities. Could it be possible in a small enough device to generate an optical wave using an alternating current at 10^{14} Hz (a wavelength of 3 microns)? However, thus far the fast evolving precision machining capability has received much more attention. In this chapter we shall briefly discuss prospects which are emerging in both these conceptual domains.

Photonics also facilitates a range of new and important applications influencing not only communication networks but also, environmental

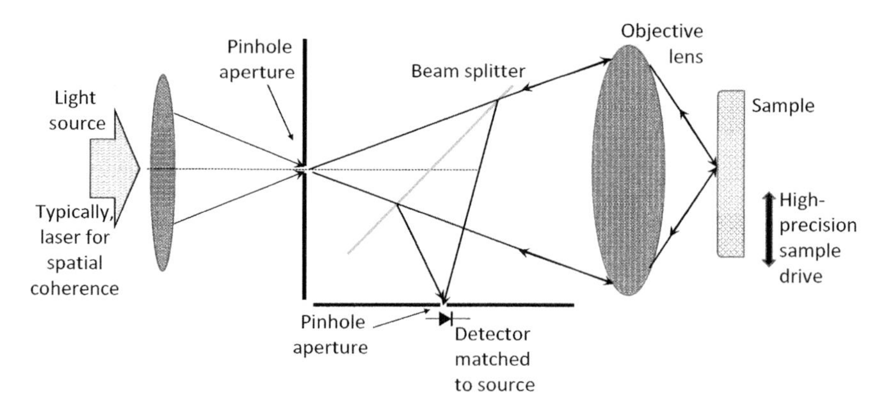

Figure 6.1 The basic elements of confocal microscopy, giving a high-resolution point-by-point scan of the sample.

characterisation and medical diagnostics and therapy. To complete the exploration of the future potential of photonics we shall also speculate briefly on where photonics could contribute beyond its current domains.

6.1 Super-Resolution Imaging

The conventional microscope is limited in its angular resolution capability by the optical wavelength and the numerical aperture of the objective lens: the limit in angular resolution is of the order of the optical wavelength multiplied by the numerical aperture (or, alternatively, in photographic terms, the wavelength is divided by the f number of the lens). In practice, realising this limit is difficult, owing to the effects of non-ideal illumination and depth of field, which combine to produce a blurred image. The middle of the twentieth century saw the arrival of confocal microscopy (Figure 6.1), which produced substantial improvements by, in effect, placing pin holes in both the illumination system and the imaging system in order to obtain much higher resolution. However, Figure 6.1 also demonstrates that the image has to be collected point by point, with the implicit need for high mechanical precision scanning of the sample with respect to the optical system. Additionally, even at its best, fundamentally this cannot break Abbe's diffraction limit on imaging resolution, which states that the maximum angular resolution is well approximated as a/λ, where a is the imaging aperture and λ is the wavelength.

The implicit assumption in this imaging limit is that the object, when viewed through the imaging system, can only be reconstructed as a far-field image; as

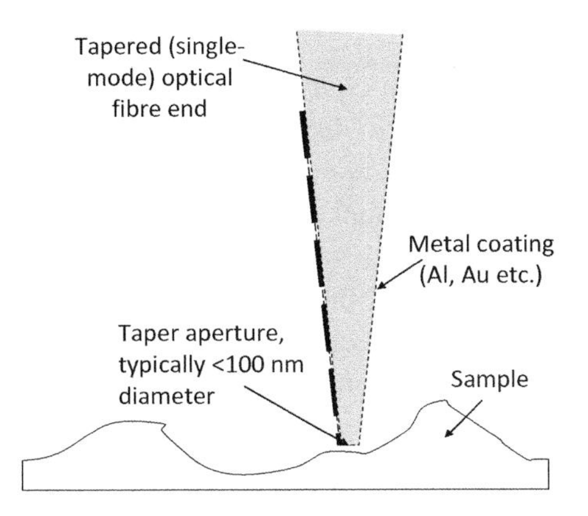

Figure 6.2 The scanning near-field optical microscope, which has a tapered single-mode optical fibre with metal coating as the illuminating source to ensure near-field operation. The tip maintains a constant subwavelength separation from the sample, using a drive similar to that in an atomic-force microscope (see the main text).

a general observation, the eyepieces in microscopes and telescopes all project the resultant image to infinity. If, however, the image could be constructed from the near field, gathering both the phase and amplitude of the illumination reflected from the object simultaneously point by point, then this restriction no longer applies. This basic idea of using the near field has been used for decades in, for example, measurement probes for electronic circuits, which plot an alternating signal passing through a circuit as a function of position (i.e. point by point) within it. The same concept has evolved into near-field scanning microscopy in the optical domain.

The concept is shown in Figure 6.2. The optical-field probe tip confines the light injected from an optical fibre to a very small, carefully defined, area. It does this by using a metal coating designed to optimise the trade-off between the inevitable high lossed incurred as light travells along a metallic path and the degree of light confinement that can be achieved. This fibre probe typically both illuminates the object and also collects the locally scattered light field. The coatings are typically gold, to minimise losses, and the probe's field of view can be confined to a few tens of nanometres. Such a probe is usually cylindrical, following the symmetry of the fibre, but some modifications, as found in, for example, the slit probe can offer specific benefits for particular applications.

This type of probe needs to make contact with, or be a consistent, very short, distance away from, the surface of the sample – it is a direct analogue of the electrical circuit forerunner mentioned above. However, the positioning systems for the fibre probe need to be very precise mechanically, to either maintain consistent contact with or, by far the more commonly encountered, to maintain a stable (to nanometres) distance from the sample.

It is here that another precision microscope technology – the atomic force microscope (AFM) system – makes its contribution. The probe from the AFM is modified to hold the optical probe and maintain a highly precise, tens of nanometres, spacing between the probe and the surface. This facilitates a local pick up for the optical field and thus generates what has become referred to as a super-resolution image. The system demands precision fabrication of the probe itself, especially if it departs from cylindrical symmetry. Achieving the necessary precision of the measurement system to determine the location of the drive network and the precision needed within the scanning x, y and z stages themselves presents challenges. At present, the use of these instruments is confined to specialist applications though there are prospects for applying the basic ideas using simpler and more adaptive probes. One example at the time of writing is the ezAFM$^+$ from NanoMagnetics Instruments.

An alternative approach to achieving effective super-resolution imaging is to treat the sample in such a way that the areas of interest have a highly non-linear reaction to the illuminating light. Possibly the most reported example of this concerns fluorescent imaging in biological systems using a cell-selective label. In this case a somewhat modified version of confocal spectroscopy, in which a pinhole is used to block out-of-focus light and so improve the image, can achieve a resolution exceeding the diffraction limit by looking at the fluorescence induced through the scanning illumination and by knowing, from the diffraction pattern, the shape of the confocal spot. In principle, then, a biological cell with dimensions in microns, which is visible using a normal microscope, can be treated with a fluorescent marker; then, under suitable illumination and precision scanning, the exact location of this marker can be determined to accuracies limited by the non-linearity of the transfer function between the illumination and the consequent fluorescence. Typically, this corresponds to 20% or thereabouts of an optical wavelength.

This discussion has merely touched on the topic of high-precision imaging using light. There are numerous other techniques, some requiring well-defined but complex illumination functions and some requiring advanced deconvolution mathematics (the received image is a convolution of the structure with the resolution profile of the system), and there are numerous other labelling techniques. Currently, all these methods are complex and they all rely upon precision scanning drives to provide location information. The basic principles are, however, well rehearsed, so speculation for the future of high-resolution

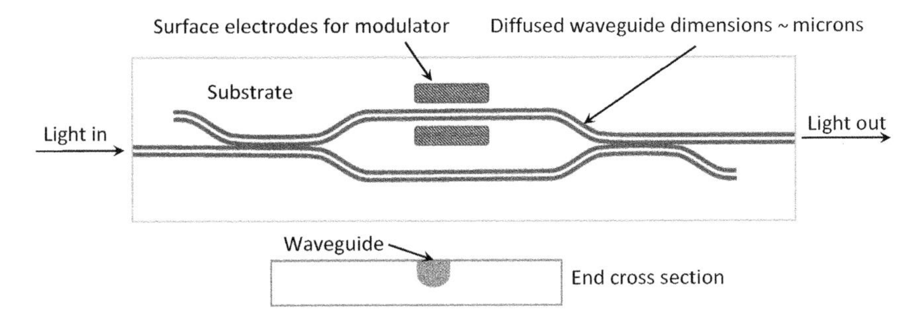

Figure 6.3 An integrated-optics diffused-waveguide Mach–Zhender interferometric modulator. The waveguide widths are of the order of a few microns.

imaging is centred around the prospects for making the technology more readily accessible to a wider audience. There are obvious uses for such technologies, especially in biomedicine, so there is good reason to be optimistic that they will evolve. For example, three-dimensional printing approaches to fabricating the probes are already emerging.

6.2 Photonic Integrated Circuits

The concepts for 'optics on a chip' emerged in the mid twentieth century, as silicon integrated circuits became a real prospect. At this stage, the preferred material, lithium niobate, was expensive and difficult to process reliably. However, demonstration optical circuits with and without modulators (which modified the material by the electro-optic effect; see Figure 6.3) showed intriguing functionality in the visible and in the very near infrared – the region which, at the time, was of greatest interest. It was in this era that one of the pioneers observed that 'integrated optics has been, is now and always will be, the technology of the future', because of the idiosyncrasies of lithium niobate processing, the need to have it in crystalline form and its intrinsic incompatibility with the formation of detectors or optical sources on substrate. Additionally, silicon was getting all the attention and all the investment.

The benefits of large-scale integration in silicon are self-evident in everyone's laptop computer and mobile phone. Silicon can also potentially act as a very good optical waveguide material, since most present-day photonics operates at wavelengths longer than the 1.1 μm cut-off for guided waves in silicon.

Silicon has, however, other benefits too – it can be machined using etching processes and photolithography into a rich variety of optically interesting shapes, from simple artefacts such as V-grooves for aligning optical fibres

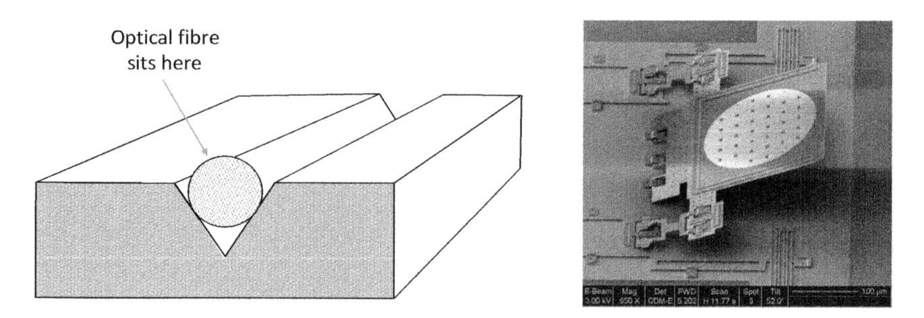

Figure 6.4 Some simple silicon micro-optics; on the left a V-groove alignment system for fibres and on the right a rotatable mirror approximately 100 μm in diameter.

with waveguides to much more complex structures such as rotatable mirrors (Figure 6.4). Silicon has the huge benefit of being transparent in the intrinsically very important 1.5 μm region, and consequently the late twentieth century saw increasing interest in silicon as the vehicle for photonic integrated circuits. The inherent versatility and straightforward availability of the basic silicon processing platform subsequently triggered a diversity of integration-compatible optical components. Waveguides can be integrated onto the substrate using a vertical etching technique, resulting in a guide lying upon the surface. Wavelength filters can be realised using a variety of grating-based approaches. Splitters and combiners are relatively straightforward extensions of the basic waveguide structure. Finally, modulators can be implemented using various forms of free-carrier modulation which influence both the optical loss and the optical phase of light passing through the modulation region. Using this principle, modulation rates approaching 100 Gbits/s have been reported (Figure 6.5).

This inevitably all applies in the near infrared band, where silicon is transparent; additionally, the question of sources and detectors compatible with this impressive interconnect facility needs to be addressed. For the sources, there are two basic approaches. The simplest and best established is to combine appropriate III–V semiconductor lasers on a silicon substrate in a suitable geometry in order to couple light into the silicon waveguide. These 'hybrid silicon lasers' receive much attention and are proving to be an effective source.

For detectors the above III–V approach is also valid, but silicon also has a crystalline structure that is geometrically compatible with germanium, so the two can be grown together to facilitate a fully integrated detector. These ideas, illustrated schematically in Figure 6.6, in principle allow a full system

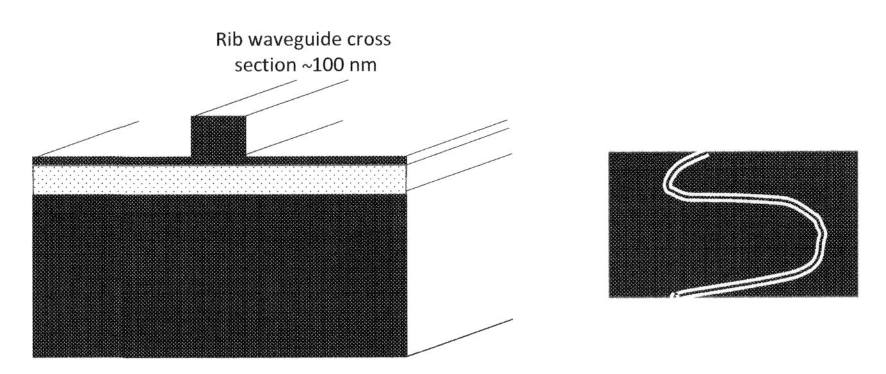

Figure 6.5 Silicon integrated optics exemplified in (left) a rib waveguide and (right) the tight bends which a high-index difference permits.

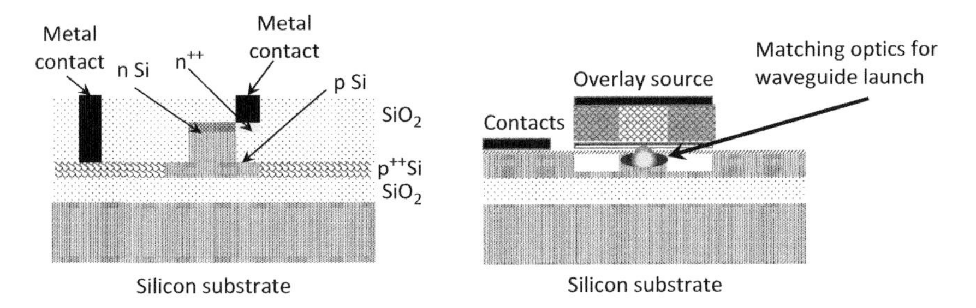

Figure 6.6 Some sample silicon circuits including (left) a modulator and (right) an integrated source.

component set on a chip. Also, this chip can be shared with all the necessary electronic drives to enable versatile, integrated, optoelectronics platform which is compatible with the essential elements of CMOS (complementary metal-oxide semiconductor) processing. At the time of writing this it has yet to become a mass production item, but the signs are definitely there for photonic integrated circuits to emerge as a 'technology of the present' relatively soon. Custom-built systems for specialised applications are already emerging.

6.3 Nanophotonics and Plasmonics

Nanophotonics is another example of emerging photonic tools and techniques which rely upon the fabrication of subwavelength structures and also on an understanding of the behaviour of light within these structures. Plasmonics is an integral part of this since it is the study of the behaviour of light within

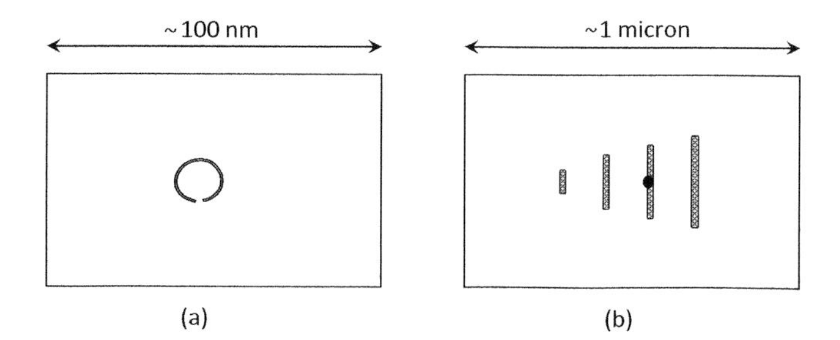

Figure 6.7 Examples of simple nanophotonic circuits including (a) a micro-resonator, where the gap is roughly equivalent to a capacitor and the ring itself to an inductor, and (b) a Yagi array antenna for use at optical frequencies.

metallic structures with these dimensions. For metallic structures the concept of 'conductivity' is well understood, and plasmonics is essentially this concept applied to optical frequencies but recognising that the free electrons in a metal can change their behaviour as the plasma resonance is approached. Nanophotonics and plasmonics are closely interlinked and have been described, by at least one of the pioneers in the field, as transferring electrical circuits into the photonic frequency range, though with due recognition of the differences in electrical properties at these very much higher frequencies.

There have been many demonstrations of nanophotonic circuits, a couple of which are shown in Figure 6.7 illustrating a ring gap resonator, which operates very similarly to an inductance and capacitor, and a Yagi antenna array, which is very close conceptually to the ultra high-frequency TV receiver antenna and indeed performs in a similar way.

There are other features of nanophotonic circuits which are broadly analogous to their much lower-frequency counterparts. Perhaps the most useful of these is the exploitation of metals as electric field enhancers at a metal – dielectric interface, particularly for pointed metallic structures. An optical 'lightening conductor'! This concept has been used for many years, particularly in the Kretchmann configuration as a sensitive probe for the dielectric constant of liquids (Figure 6.8(a)). This basic concept has also been coupled into photonic integrated circuits illuminated by a fibre waveguide (Figure 6.8(b)). The generic idea has also been used to enhance the electric field in solar cells and significantly increase the efficiency of some biophotonic therapies.

Nanophotonics and plasmonics currently attract widespread and enthusiastic research endeavours, since even this brief discussion should have indicated their remarkable potential. In common with all the topics mentioned in this chapter there are textbooks and journals dedicated to these topics.

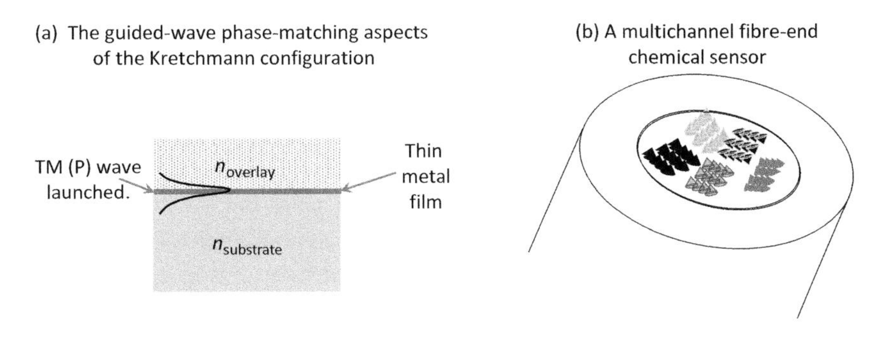

Figure 6.8 A nanophotonic field-enhancement system (a) in the guided-wave traditional Kretchmann configuration. The launched light must phase-match precisely into the propagating light, with the field distribution indicated. (b) Metallic prisms on the end of a fibre, with different segments each targeting a separate chemical species on a probe just a few microns in diameter. The electric-field-enhancement properties of the nanophotonic metallic islands are critical to its operation.

6.4 Metamaterials

A metamaterial (mentioned briefly in Chapter 4) is conceptually a material ensemble with structural dimensions far below the wavelength of the electromagnetic (or indeed acoustic) fields with which it interacts. Through this interaction, structural variations in the metamaterial can produce a response that would otherwise be unobtainable. Whilst structures such as gratings could be argued as falling within this classification, a moment's reflection demonstrates that the optical properties of these structures are dictated by the properties of their component materials and the way in which these component materials are organised. The component materials behave in such structures exactly identically to their bulk equivalents. Metamaterials, in contrast, are made-to-measure three-dimensional materials with structural variations in dimensions considerably less than a wavelength, resulting in an overall material which behaves as a bulk system with entirely different properties. The incoming optical wavelength is too long to 'see' the structural artefacts!

Typically, metamaterials can comprise arrays of nanophotonic circuits (Figure 6.7) with resulting properties dictated by this structure (resonance, for example). Alternatively, arrays of other material structures can be arranged in such a way that the incident light beam 'sees' the average properties of the structure (i.e. any structural variations must be well controlled and extremely small compared to a wavelength). Another possibility is arrays of quantum dots, where the optical properties of the materials themselves are modified by dimensional features small enough to affect the quantum mechanical energy levels within the structure. And all these ideas – and others – can in principle be combined to synthesise almost any conceivable material response!

These brief qualitative comments already point to intriguing prospects in terms of manipulating light transmitted into a metamaterial, and therefore open up opportunities to implement previously intractable functionality. Perfect imaging and cloaking are possibly the most discussed examples.

Perfect imaging thwarts the Abbé resolution criterion by utilising a material with a refractive index of -1 to produce the imaging function. Assuming that the material is perfectly homogeneous from the perspective of the light passing through it and that it has perfectly parallel sides and perfectly flat surfaces, the ray diagram, as shown previously (Figure 4.17), demonstrates paths through which a perfect image is formed. It images using 'rays' rather than 'waves'! The imaging material also has to be thicker than the separation of the material and the object. Potentially there is the scope for effective magnification through making ever thicker samples of a material with index -1. From a 'Clausius–Mossotti' perspective, a negative index can be achieved through configuring a synthetic molecule-like structure which reradiates the input as an antiphase output – remember that is the context of refractive index it is phase velocities rather than group velocities which dictate what happens. There has been some practical progress in realising negative-index optical materials (though not as yet the super-lens envisaged as a lens composed of material with index -1), but this has been achieved through nanophotonic resonators evolving from the original electromagnetic demonstrations at microwave frequencies. There are then practical constraints since the resonators function over only a very narrow spectral range and need to be extremely precisely matched for full efficacy. The super-lens, however, continues to fascinate.

A 'cloak' renders the object which its surrounds totally invisible (Figure 6.9). The necessary index distribution to ensure that this actually happens can, in principle, be readily calculated for a particular candidate object for cloaking. In practice, such a distribution is tractable only for simple objects, for example cylinders or spheres. Again the nanophotonic resonator appears as a key component so cloaking for all colours of the rainbow clearly presents issues, because resonance only occurs for one wavelength. The cloaking principle has, however, been successfully demonstrated, most notably at microwave frequencies where the fabrication tolerances are four or more orders of magnitude less demanding than at optical frequencies. The invisibility cloak has an instant and easily communicated appeal and continues to fascinate the photonics community, not to mention the world at large!

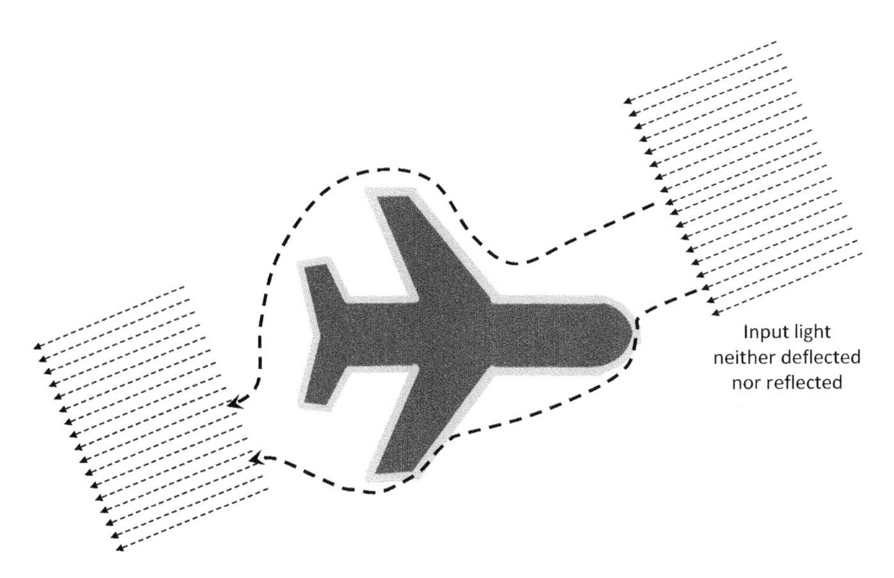

Figure 6.9 How cloaking might work – but broadband cloaking, even for simple shapes, and full cloaking for complex shapes have yet to be realised.

6.5 Photonics and Materials – A Brief Summary

In the preceding sections we have briefly discussed super-resolution imaging, photonic integrated circuits, plasmonics and nanophotonics and metamaterials, all of which are establishing themselves as essential building blocks for future photonic systems and their applications. At a conceptual level, they all rely upon three essential building blocks:

- A thorough understanding of optical materials and their properties at photonic frequencies. This also includes the interfaces between different material components and the variations in interface behaviour with the overall structural dimensions of a particular material with respect to the wavelength.
- A thorough understanding of the behaviour of ensembles, particularly those on a regular repeating pattern in two and three dimensions, again with due attention to the ensemble component sizes with respect to the optical wavelength. Effectively this is a generalisation of the Clausius–Massotti equation, which relates what is absorbed by to what emerges from a component after excitation by an electro magnetic wave.
- A full appreciation of the effects of structural geometry within a sample of material comprising the synthetic optically active elements mentioned above. In particular, this applies to three-dimensional regular periodic arrays with specified quasi-particles on each of the array nodes.

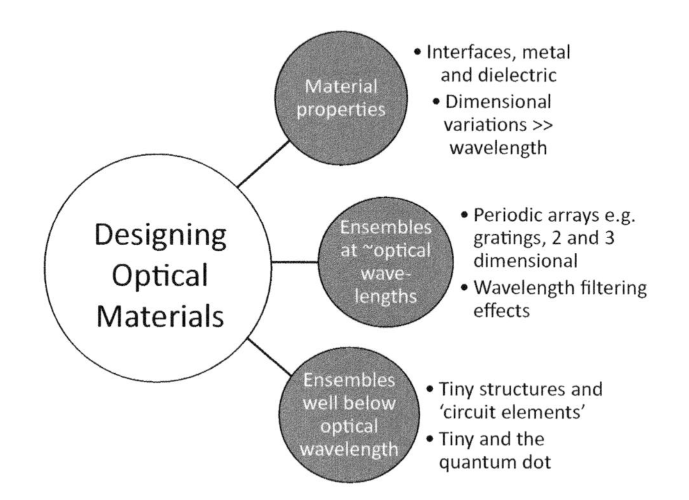

Figure 6.10 The building blocks for designing optical materials and the essential options dependent upon critical scale.

These three conceptual building blocks (Figure 6.10) are common to any situation in which waves interact with materials, whether the waves be electromagnetic, mechanical, acoustic or quantum mechanical. They are at the heart of much of our scientific and technological evolution.

6.6 Entangled Photons

The concept of 'entanglement' consolidated around the turn of the present century. At a basic level, the concept is relatively straightforward. In this book we have assumed that the reader is comfortable with the idea that light travels as a wave and yet is detected as a particle, the photon. Consequently, if we can build up an image photon by photon in, for example, the Young's slits interference experiment, then each photon emitted from the source would arrive in the detection plane at a position whose statistical probability is determined by the interference pattern. This would, in turn, reflect the wave-defined multidirectional path through which a particular photon trajectory happens to proceed.

Thus, much endeavour in the optical physics community has been directed towards realising controlled single-photon sources. There has been some success in this pursuit already. It opens up at least two intriguing possibilities. First, if a single-photon source can be realised, then by implication a single-photon detector can also be achieved, and indeed this has been achived: the single-photon avalanche detector (SPAD) is now a catalogue item. This enables the detection, and therefore the characterisation, of single photons emitted from an illuminated object. There are prospects here in medical imaging through tissues in order to detect the position of a catheter probe, by including a fibre source in the catheter and using the arrival of the first photon to reach the SPAD to indicate the position of the catheter. This assumes that at least one photon can penetrate through the tissue without being scattered, an assumption may be plausible but has yet to be definitely proven. The second possibility concerns new-generation communications networks, where generating and later detecting a single photon can contribute towards establishing reliable connections.

The concept of 'entanglement' takes this somewhat further. There are situations where an atom can generate two photons in a closely linked process and therefore the properties of these two photons are intimately related, even permanently bound together. Similar arguments can be extended to other elementary particles when closely bound together through some single atomic or nuclear process.

The intriguing features of these two (or indeed more) 'entangled' particles is that after entanglement, even if the individual particles become separated or take different 'wave paths' to their destinations, their overall net properties will remain the same and they will need to 'recognise each other' in order to be detected again. Consequently, perhaps a sequence of such entangled photons could be formed to represent a coded message. The components of these entangled photons would then be separated from each other in some way and the encoded message could only be read when the two sets of components come together again. This is the essential thought process behind quantum communications and unbreakable code sequences, a generic area of interest which has become increasingly important as the internet rapidly expands, and security becomes increasingly critical. The recipient of the code would need to have both components to hand; this would prevent any interception.

The quantum computer is another intriguing potential long-term manifestation of these ideas, though, as yet, it has only been demonstrated in a very simple form. In contrast, entangled-photon communication systems were demonstrated at the beginning of the twenty-first century and have been gradually improved since then. However, the systems remain many orders of magnitude slower than their conventional fibre-communication forefathers but both the science and technology continue to progress leading to an, as yet undefined, probably currently unknown, final form.

The entangled-photon concept has also asserted itself as potentially the ultimate computational resource, and simple logic functions have been demonstrated. For example, an entangled communication system can be interpreted as a form of AND gate, which only gives an output when identical signals and codes coincide in space and time.

Entanglement will continue to fascinate for a considerable time to come whether for photons or other elementary particles. Quite how, when and for what it will emerge into the outside world remain intangible questions. The concepts do appear in advanced physics textbooks, but to date perhaps the most accessible version of the story is in Jonathan Dowling's book, mentioned in the Further Reading section at the end of the book. This very readable text tells the story for 400 pages without a single equation though it does contain conceptually demanding and thought provoking accounts of the basic ideas and their possible applications.

6.7 Graphene and Other Exotic Materials

The ever expanding capabilities to visualise, fabricate and characterise material structures with dimensions measured in nanometres – much below the wavelength of light – has facilitated numerous intriguing innovations some of which will undoubtedly find their way into engineering applications. Of these graphene (Figure 6.11) is among the most intriguing. This single-layer crystalline

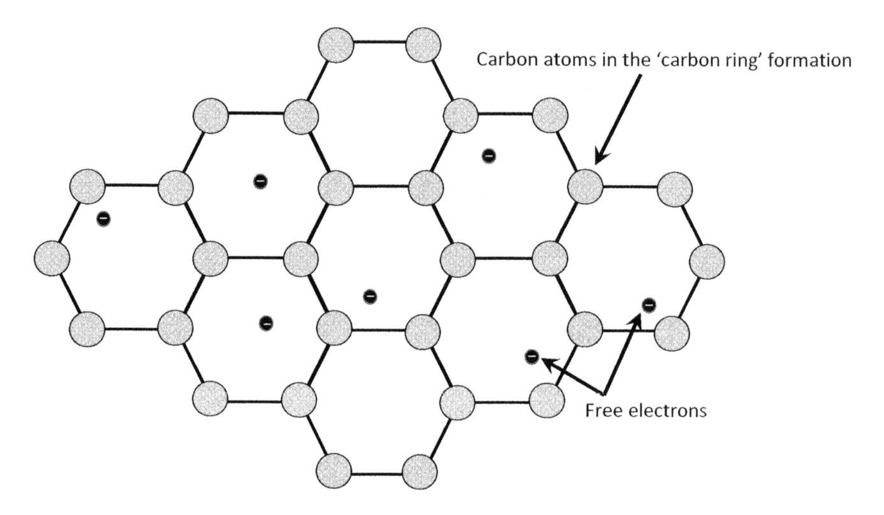

Figure 6.11 Illustrating graphene: a single layer of carbon atoms linked together with free electrons flowing within the layer. The atoms are about 0.14 nm apart.

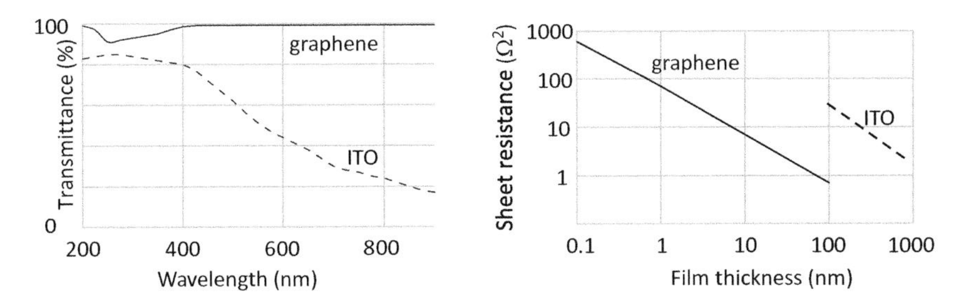

Figure 6.12 The properties of graphene in comparison with the currently preferred transparent contact material, indium tin oxide (ITO).

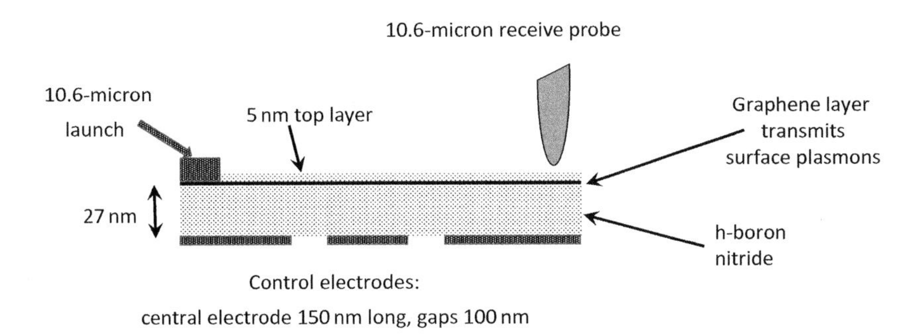

Figure 6.13 A nanostructure system based on graphene giving a voltage controlled phase shift of up to ~180° at 10.6 μm wavelength.

carbon has many interesting features. It has high electrical conductivity, it is mechanically strong whilst also being flexible and it is uniquely transparent across the visible spectrum and beyond (Figure 6.12). In the photonics context, graphene is obviously a candidate for touch-screen displays particularly where highly flexible screens are required.

There is much more to thin layers of graphene than their action as conducting electrodes and many intriguing optical–electronic interactions have been demonstrated. Figure 6.13 shows an interesting example where a tiny structure, nanometres in dimensions, is sufficient to produce a full 180° voltage controlled phase shift for a 10.6 μm laser wavelength. The diagram also illustrates the use of a nanodimensional probe to receive the modulated beam. (This may even be viewed as a slightly eccentric mid-infrared transistor!) The science and application of graphene continues to excite and intrigue – there is little doubt that its engineering potential will eventually be realised.

The structure in Figure 6.13 also involves another intriguing material; h-BN, hexagonal boron nitride. This has a very similar layer-by-layer crystalline structure to graphene but conveniently has a large bandgap (over 5 electron volts) and a relatively low electrical conductivity, so its action is to preserve the plasmon (i.e. the optical-frequency electric current) within the graphene layer. This is another example of novel material exploitation in the photonic domain facilitated by these ever improving fabrication and manipulation technologies.

6.8 Manipulating Light Sources

There are other innovations in photonics which promise to facilitate significant advances in performance. All involve in some way manipulating light with materials or, strictly speaking, material structures, since it is increasingly the combination material properties and structural properties which facilitates the manipulation of optical signals. Squeezed light is but one example. The discussion of noise in the context of photonics historically centred around the concept of 'shot noise', which we have already explored; it represents the random change in the rate of arrival of photons. We have also mentioned that this can be viewed in the 'analogue' domain, with the wave represented by a vector in which the shot noise impacts equally on the amplitude and phase. However, if a means can be found to turn the phase fluctuations into amplitude fluctuations, or vice versa, then beating the shot noise limit by squeezing the light (to reduce phase noise) becomes a real possibility. This assumes that after the squeezing has been completed, the detection processes that follow are somehow made insensitive to the noise increase incurred as a result of the process.

Whilst this technique has been already utilised in the Laser Interferometer Gravitational-Wave Observatory (LIGO) as mentioned in the previous chapter, the implementation in the gravitational-wave interferometer was optically bulky and extremely expensive. Squeezed light, however, becomes more interesting when it can, in principle, be combined with guided wave optics and produced using nanometre-precision machined devices. (Figure 6.14). Thus far examples have been based on using subtle non-linear interactions or even micro (strictly nano) mechanical resonators, in both cases to inhibit phase fluctuations at the expense of enhancing the intensity variations. The extent of the squeezing using these devices has thus far been modest – one or two decibels. The integrated circuit demonstrations sometimes also involved cryogenic cooling, so there remains much scope for improvement towards the LIGO exemplar of well over 10 dB. The ideas of squeezed light emerged in

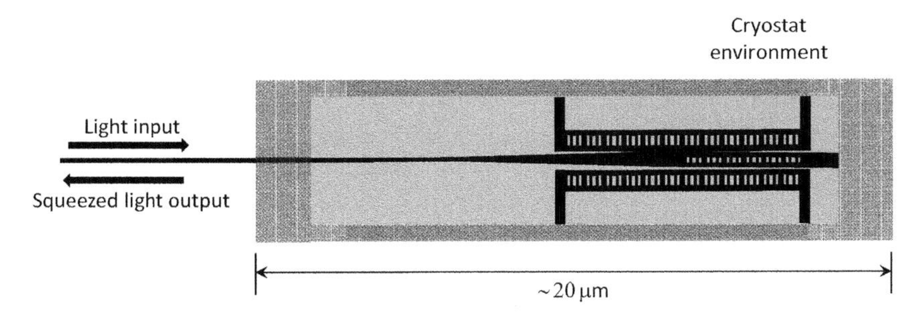

Figure 6.14 An exploratory mechanical structure for squeezing light in the phase domain, based upon nanoscale micromechanical resonators. The light grey area is a vacuum enclosure. The resonator, shown in black, is freely mounted over a cavity in the substrate structure. The striped element coming from the right-hand side is an optically excited resonator. The long striped rectangles facilitate coupling between the light and the mechanical resonator.

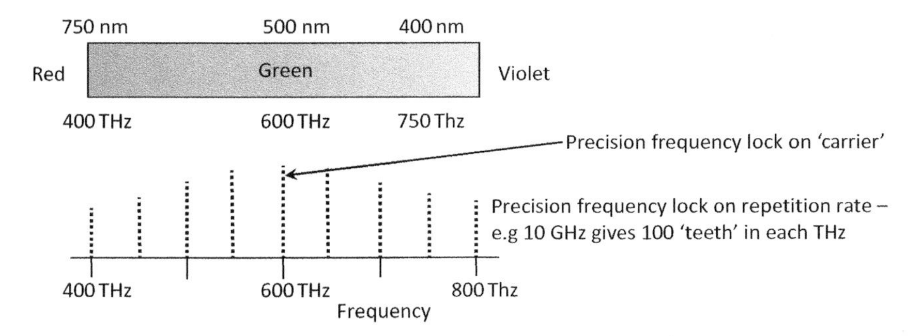

Figure 6.15 The basics of the frequency comb utilising a 2.5 femtosecond pulse in the green and, in this example, based upon a 50 terahertz pulse repetition rate. More typically, a 1 Ghz repetition rate would give 400,000 'teeth' over the same bandwidth, in principle enabling accurately controlled spectroscopic measurements.

the 1980s but the thought of beating the shot noise limit in a compact and cost efficient format remains a significant challenge and a source of great interest.

A frequency comb is another example of an intriguing optical source. The concept originated as a means towards high-precision spectroscopy, where the source is locked through some form of multiplier to a radio frequency master oscillator (Figure 6.15). The output from such a system when viewed in the time rather than the frequency domain is a sequence of very short pulses separated by the master frequency. The exact techniques through which such 'combs' can be generated involve ultra-precision non-linear optics and opto-electronic design. The special feature of the frequency comb is that each component is accurately phase locked with respect to the others, resulting, in the time domain, in unrivalled phase noise performance among the various pulses within the pulse train. The applications of this continue to be explored: optical clocks using frequency combs have been demonstrated and highly precise spectroscopy, particularly in the time domain, is another area of exploration. Just as in squeezing, a few tentative demonstrations of combs realised in photonic integrated circuits have begun to emerge. In common with squeezing, frequency combs continue to attract much attention.

6.9 Summary

The aim of this final chapter has been to give some insight, albeit superficial and far from complete, into the immense potential which photonics offers for the future. There are continually evolving new photonic tools. In parallel, continuing new societal challenges emerge for which the tools offered through photonics promise innovative and effective solutions. Figure 6.16 attempts to encapsulate at least some of this potential on the basis of our discourse throughout the book and the remarkably consistent recognition among numerous international organisations of the challenges which society currently needs to address. There are other indictors too, including, for example, the fact that, at the time of writing (2018) the European production output in photonics exceeded that in electronics and the research and development endeavour within this sector continually gains more momentum. There is indeed some confidence in asserting that 'photonics is the electronics of the twenty-first century'.

- Make solar energy economical
- Manage the nitrogen cycle
- Advance health informatics

- Prevent nuclear terror

- Advance personalised learning
- Provide energy from fusion
- Provide access to clean water

- Engineer better medicines
- Secure cyberspace
- Engineer the tools of scientific discovery
- Develop carbon sequestration methods
- Restore and improve urban infrastructure
- Reverse engineer the brain
- Enhance virtual reality

Figure 6.16 A view of the technological challenges currently facing society and the areas to which photonics promises to contribute (*based on a US National Academy of Engineering 2009 list, still current, and similar to the listings for many professional organisations, e.g. Photonics 21 in the EU, the National Photonics Initiative in the US etc.*). (The prevention of nuclear terror is separately boxed because photonics is unlikely to make a direct contribution to this large-scale technological challenge.)

6.10 Problems

Problems involving predicting the future are by their very nature encroaching on the totally inaccessible! Answers will be educated guesswork and will vary with time and among communities. The whole 'future scenario' discourse is one to speculate about with fellow enthusiasts. So the problems here are in two categories: in section A, the reader is invited to look in a little more depth at a selection of current ideas within the community and in section B there is a discursive section inviting speculation on what might happen and why.

Section A

1. (a) How could a Mach–Zehnder interferometer in integrated optic format become an intensity modulator (Figure 6.3)? Remember the electro-optic effect and note that the early devices, of which this interferometer is an example, used lithium niobate as the substrate.
   (b) What parameters in this device could limit its modulation bandwidth and its possible modulation depth? How might this change if your final system application were the design of a phase or frequency modulator?
2. (a) Figure 6.6 shows some silicon integrated optics. How does the modulator (on the left-hand side of the figure) function in this particular case?
   (b) When compared to the device in Figure 6.3 what might its limitations be in terms of operational optical wavelength range and in achievable phase and intensity modulation depths?
3. (a) Figure 6.8(a) shows the field distribution for a wave propagating along a dielectric–metal–dielectric interface. Suppose that the lower layer (substrate) is one face of an equilateral prism and light is incident on one of the other faces. How does the incident angle on this face required to produce propagation along the metallic surface depend on the propagation constant of the surface plasmon wave? How might this change if the overlay index and thickness were varied? The clue lies in phase matching at the interface. This prism arrangement is the basis of the Kretchmann configuration used in overlay index measurements.
   (b) why does this arrangement need a TM wave to be launched and what would happen to a TE wave? (In a transverse magnetic (TM) wave the electric field oscillations are perpendicular to the interface under consideration, and in a transverse electric (TE) wave they are paralled to the plane of the interface.) The system shown in Figure 6.8(b) has no facility for phase-matched launching. Is this important and if so (or, indeed, if not), why?

4. Figure 6.13 shows a phase modulator which has been demonstrated for the 10 micron wavelength region. Given the dimensions in the figure, and given the observation that the signal travels from the launch electrode input to the receive probe as a surface plasmon (i.e. an electric current), comment on how the modulator modifies the phase delay and on any implications this might have for the effective electron velocities in the graphene layer. Figure 6.12 also contains some pertinent background.
5. Figure 6.14 shows a mechanical structure fabricated from silicon and designed to perform optical squeezing, in order to reduce the phase noise below the shot noise limit, albeit in practice to a modest extent. Consider how this precisely micro-machined structure works Why does it need to operate in a cryostat? How would you view the future prospects for such a concept?

Section B

6. There are many respected sources of information overviewing interesting technical developments in photonics (Photonics 21, US National Photonics Initiative and several others, including the professional societies in optics such as SPIE and OSA). Look for the latest freely available reports on trends from these organisations and compare and contrast these predicted trends for agreements and differences. How might these trends map into the societal needs identified by numerous other professional organisations such as the National Academies of Engineering, the latest EU research programmes and socio-technological organisations such as the Royal Society of Arts, UK?
7. Use collective imagination and insight (remembering that fresh views are very valuable!) to consider other feasible photonics initiatives, such as combining parallels from the bio-sciences, materials science, very-high-precision machining, quantum systems, electronics and information availability, for example.
8. Speculate on the detail of a few future technologically possible applications suggested in Figure 6.16. Technological feasibility is, however, but one factor influencing the eventual realisation of a new technology. Community acceptance is perhaps the most important of these. Will people use it and be willing to pay? What factors may come into this in the specific cases you have chosen? Considerations on safety, and standards for demonstrating safety, performance repeatability, rivalry with established practices aimed at performing a similar function and many similar factors come into play. Look up Porter's five forces in this context.

Appendix 1
More About the Polarisation of Waves

The basics of wave polarisation are very straightforward. When a polarisation state is referred to as being linear it is simply an electromagnetic wave whose electric field is vibrating in one plane, as shown diagrammatically in Figure A1.1(a). Conventionally the electric vector defines the wave, and in the figure the direction of polarisation is vertical. Figure A1.1(b) shows two waves in phase with each other and of equal amplitude but one with the electric vector horizontal and the other with the electric vector vertical, as in (a). Since both these waves have the same amplitude, then what would be the amplitude and polarisation state of the resultant when the two are added together?

In Figure A1.1(c) the horizontal and vertical components are 90° out of phase. How would the resultant electric vector for this particular wave combination evolve? If your eye is looking at it from the arrowhead point, how would the resultant electric vector behave as it moved towards you? The direction of rotation defines clockwise (right-handed) or anticlockwise (left-handed) polarisation states. If the two waveforms in (c) were not exactly in quadrature then how would the resultant state behave? As a hint the 'not quite in quadrature' phase differential can be resolved into an in-phase and a quadrature component. The net result is elliptical polarisation.

We have mentioned several times birefringent materials, in which different polarisation components travel at different velocities. We usually consider the case of linear components, shown diagrammatically in Figure A1.2, though some materials are naturally circularly birefringent, depending upon the molecular symmetries concerned – sugars are an example. In the linearly birefringent case in Figure A1.2, an input wave (A) oriented at 45° to the principal axis will split into slow and fast components. The input plane to the birefringent medium is shown at top left in the diagram. The phase difference between the two components increases as the propagation distance through the birefringent material increases. For example, at point (B) in the

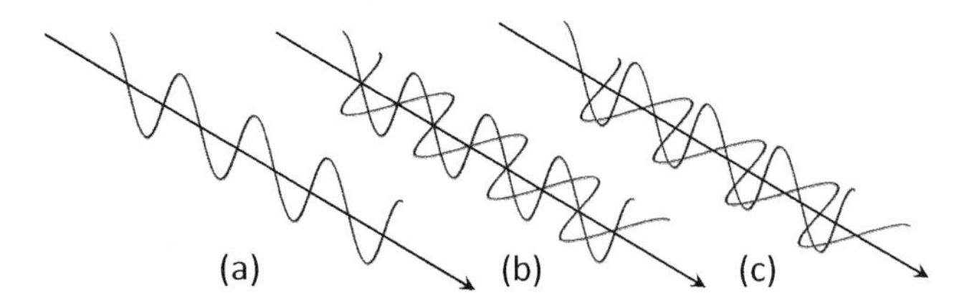

Figure A1.1 The electric field vibrations in various polarisation states: (a) a simple linear state; (b) a composite linear state with horizontal and vertical components; and (c) a circularly polarised state with vertical components shifted by 45°.

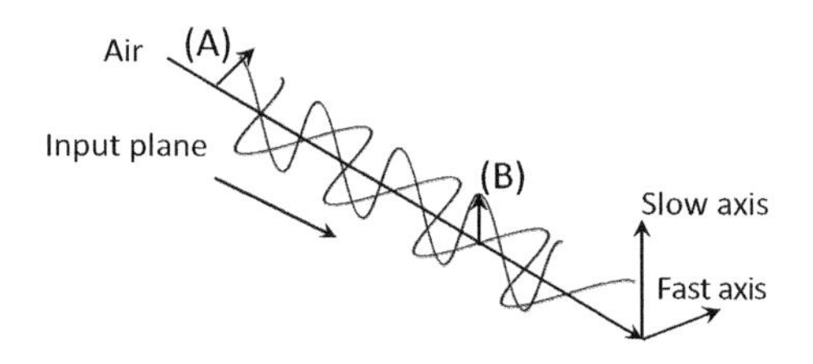

Figure A1.2 Propagation of an arbitrary input polarisation state through a birefringent medium analysed by resolving the input into states along the principal axes of the medium.

diagram the two components are in quadrature so the output from this thickness of the medium would be circularly polarised. If we were to proceed further, the two components would become 180° out of phase, and the output would be linearly polarised but orthogonal to the input at (A), and so on. Birefringent materials used in this manner – often referred to as phase plates – are extremely useful components in optical systems.

This moves us into questions concerning the influences of the input light on the output fields. As an example suppose that the distance between the point (A) and the point (B) in the diagram was kept constant but that the input frequency was doubled. Assuming no net dispersion in the behaviour of light travelling along the fast or slow axis then the output would be back to linearly polarised but at 90° to the input.

This gives a hint on how a broadband source might behave when passed through a phase plate of this nature in this particular format. The broadband source would give slightly different polarisation states at the output depending upon the wavelength of the input, and these differences will become greater with a longer path within the birefringent medium, which can be, in principle, extended as far as we wish. Thus the light ends up unpolarised. This assumes that the input state is polarised, and for sources such as super-luminescent light emitting diodes this is the case. A depolariser which operates in this manner is often useful in conjunction with such sources.

The situation gets more complex if we remove the assumption of spatial coherence – in other words if the input beam has components travelling in many directions rather than simply in one direction as in Figures A1.1 and A1.2. This is an interesting problem to ponder but in practice many situations in optics can be visualised using the approaches that we have adopted here.

Further Reading

E. Collett, *Field Guide to Polarization*, SPIE Field Guides, Vol. FG05, SPIE, 2005.

Appendix 2

The Fourier Transform Properties of Lenses

Here we aim to give an intuitive appreciation of the Fourier transform properties of a complex lens. Suppose we have (Figure A2.1) a grating with an *amplitude transmittance* (the percentage of the input amplitude which emerges after transmission through the grating) which is simply a sine wave, as indicated. We can consider any amplitude transmittance function as a sum of such sine waves through a straightforward Fourier transform relationship. Note that here we are discussing *amplitude* transmittance throughout, although most photodetectors, our eyes included, respond to intensity levels.

We will take the maximum transmittance as unity. The average amplitude transmittance is shown as a dotted line. There will be interference between amplitudes scattered from different parts of the grating; a more detailed analysis (see for example Goodman's excellent book) demonstrates that the two diffracted beams indicated in the diagram have amplitudes proportional to the amplitude of the sine wave indicated (but that this includes the 'positive' and 'negative' frequency components, on either side of the optical axis) and also that the amplitude of the transmitted undeflected wave is proportional to the mean amplitude transmittance of the grating. Furthermore, the deflection angle θ_d is given by

$$\sin\theta_d = \frac{\lambda}{d} \tag{A2.1}$$

where d is the grating spacing and λ is the wavelength of the input light. Conveniently, the quantity $1/d$ can also be referred to as the spatial frequency of the grating so, for angles where $\sin\theta_d$ can be approximated to θ_d the diffraction angle is proportional to the spatial frequency. This above approximation is perhaps more versatile than might at first be guessed – even at a diffraction angle of 45° the approximation is less than 10% in error. So we see that the amplitudes of the diffracted components are proportional to the amplitude of the spatial frequency component in the grating corresponding

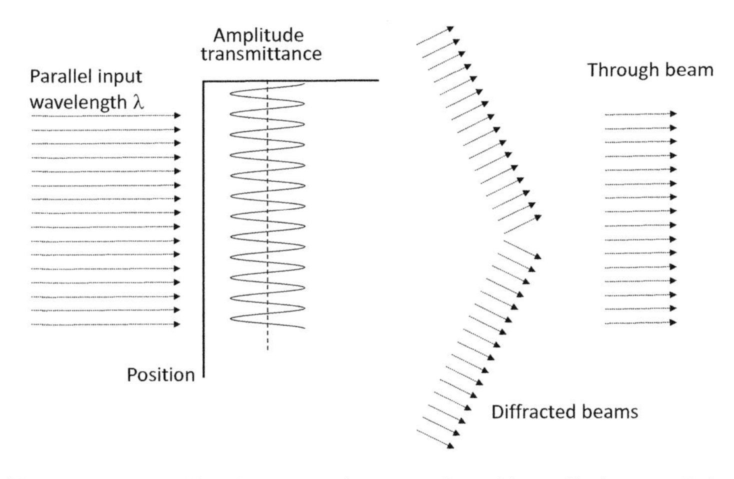

Figure A2.1 The diffraction pattern from a grating with amplitude transmission varying sinusoidally in one dimension.

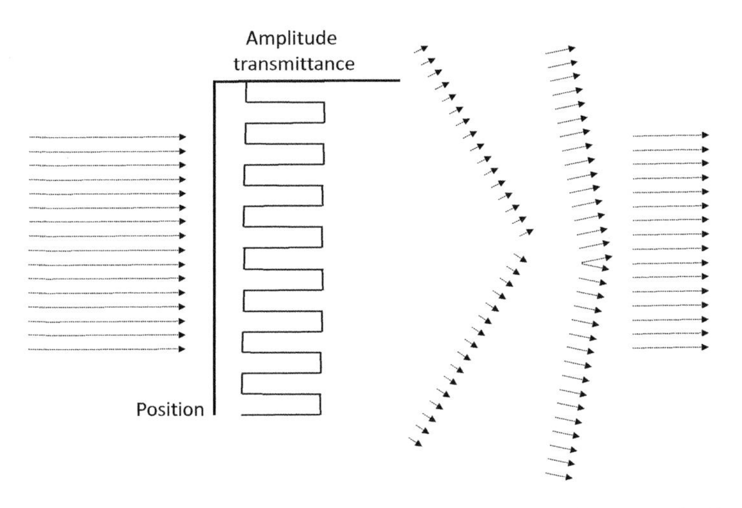

Figure A2.2 A square wave grating – only the odd harmonics appear in the diffraction pattern with amplitudes varying as the inverse of the harmonic number (diffraction order).

to that diffraction angle and that amplitude of the straight-through portion is proportional to the mean amplitude transmittance.

Moving to a specific example of a more general case, in Figure A2.2 we consider a square wave grating. Here the lowest-order diffraction angles and amplitudes correspond to the fundamental amplitude component of the transmission function for a square wave. However, a glance at the diagram shows

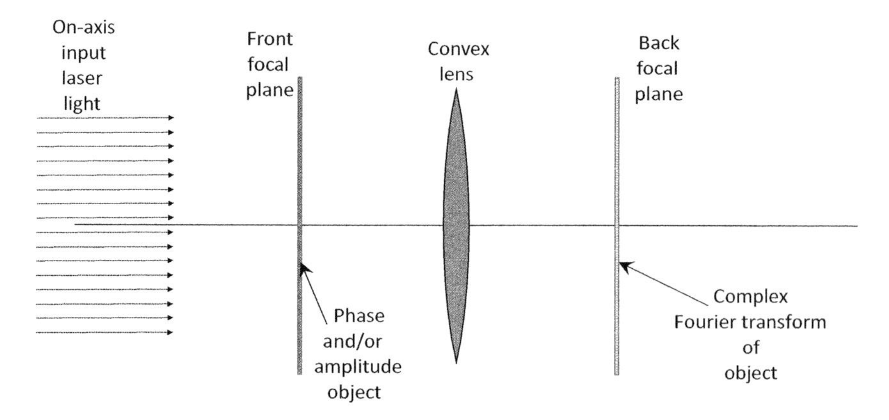

Figure A2.3 A convex lens set-up for obtaining the Fourier transform of an input object.

that for twice the angle of diffraction for the fundamental component there is no diffracted wave – square waves only contain odd harmonics. The third harmonic amplitude has one-third of the amplitude and three times the deflection (in the $\sin\theta \sim \theta$ approximation) of the first harmonic, and so on. Furthermore, the amplitudes of the odd harmonics vary inversely as the harmonic number (diffraction order).

Now imagine a convex lens collecting all this diffracted light (Figure A2.3). The back focal plane of the convex lens (neglecting distortions, aberrations and so on) will exhibit a series of dots spaced by a distance proportional to the diffraction angle (strictly speaking, the tangent of the diffraction angle is the focal length divided by the distance from the point where optical axis meets the back focal plane, but we are assuming the same $\sin\theta_d \sim \theta_d$ approximation as before). Furthermore the amplitude of these spots in the focal plane will be proportional to the amplitudes of these spatial frequency components. If the lens collects all the diffracted light, an accurate image will be produced wherever the image plane is given by the traditional lens formula. However, if the lens misses any of this light then the corresponding amplitude components will be reduced or even eliminated, so the image will have a different spatial harmonic amplitude distribution from that of the original grating. For example, if we miss the entirety of the third-order beam and above, we will be back to an sinusoidal amplitude variation in the image plane at the fundamental spatial frequency of the grating, since we now have first the average illumination and the two first-order components to make up the final image.

The final nuance comes if the input object (our input grating) is placed exactly in the front focal plane of the lens. In this case the focussed light in the back focal plane represents both the amplitude and phase components of the input amplitude transmittance. The concept of phase components in the amplitude transmittance may appear somewhat arbitrary – where is the phase reference? The simplest approach to this is to take the mean transmittance (zero deflection) as zero phase and define all the phase components with respect to that. This is also effectively the standard approach in normal Fourier transform operations and gives an insight into how an object which has only a phase transmittance and no change in amplitude can be visualised by changing the zero-frequency component of the Fourier transform in the back focal plane of the lens. The reader should examine the Fourier transform expressions for a phase-modulated carrier and the striking way in which changing the phase or even removing the fundamental components turns the input *phase* distribution into a plane *amplitude* output distribution.

Figure A2.3 shows a set-up that can be used to perform Fourier transforms on any arbitrary input object and also indicates how, by manipulating the transmission properties at the Fourier transform plane, the properties of the resulting image, formed by the lens after the Fourier transform plane, can also be modified. This concept has found enormous application in feature enhancement (e.g. in highlighting the edges in an object and thus producing an image with the lower spatial frequency components attenuated or sometimes removed) and in various forms of digital image manipulation software. For the latter, of course, the hardware between the input object's image and the final output is not there. However, in imaging systems from microscopes to telescopes these phenomena are always present and this approach to hardware-based imaging is often used. Perhaps the most familiar is 'dark ground imaging' in microscopes, which enables the visualisation of low-contrast objects, including those comprising only phase-delay detail. These phenomena determine the quality of the final image, how this image quality can vary with the illuminating wavelength (e.g. through chromatic aberration in lenses or the fact that blue light is less diffracted than red) and many other features of imaging systems.

Further Reading

J. W. Goodman, *Introduction to Fourier Optics*, McGraw Hill, 2017. First published in 1968. Remains the most respected text.

D. Voetz, *Computational Fourier Optics*, SPIE Tutorial Texts, Vol. TT89, 2011.

G. O. Reynolds *et al.*, *Tutorials in Fourier Optics*, SPIE Press, Bellingham, WA, 1989.

There is also a free IPad App which performs spatial Fourier transforms on photographs stored on the device – very instructive!

Appendix 3

The Clausius–Mossotti Equation and the Frequency Dependence of the Refractive Index

The key here lies in the concept of polarizability – in other words the modifying effect of an electric field on the charge distribution in an atom or molecule. Figure A3.1 shows that an applied field will induce a net dipole moment on the molecule and that this in turn distorts the field distribution around the polarised molecule. Consequently, in a dielectric medium the effective field on each individual molecule is the sum of the applied field and that due to the induced dipole moments within the medium – here assumed homogeneous and isotropic.

The dielectric constant or relative permittivity, ε, gives a measure of the polarisation effect. Consider the capacitor shown in Figure A3.1. By definition,

$$\varepsilon = \frac{\text{applied field}}{\text{nett field}} = \frac{C_1}{C_0} \tag{A3.1}$$

where C_1 and C_0 are respectively the capacitance with and without the dielectric present. The ratio ε can be quite large; for silicon, for example, it is nearly 12 owing to the near-cancellation of the applied field within the dielectric.

The *Clausius–Mossotti* equation relates the dielectric constant of a medium to the polarisability of its molecules. For a particular frequency within the absorption spectrum of the molecule, it can be shown that

$$\frac{\varepsilon_r - 1}{\varepsilon_r + 2} = \frac{1}{3\varepsilon_0} \sum_j N_j \alpha_j \tag{A3.2}$$

where N_j is the volume density of the molecules resonating at a particular resonant mode j. The symbol α_j is the polarizability (i.e. the induced dipole moment per unit applied field) of the molecule and is a function of frequency. The summation in the equation is over all resonances pertinent to the frequency of interest at which the value of ε_r is to be determined.

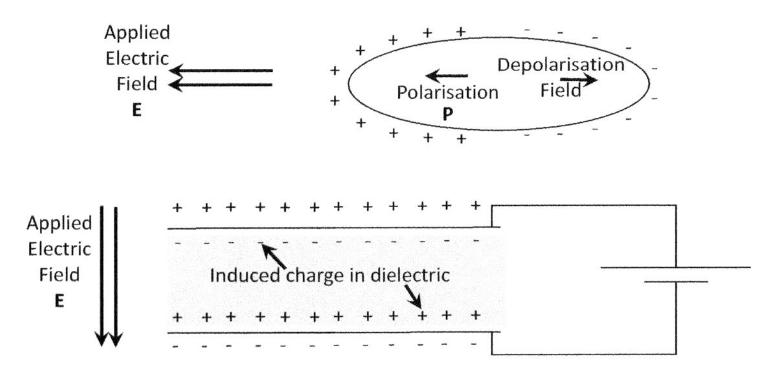

Figure A3.1 Illustrating how an applied electric field displaces the charges in a dielectric and how this produces a depolarising field that partly cancels the applied field within the dielectric.

The essential aspect of equation A3.2 is that α_j can be derived as the response, at the frequency of interest, of the 'mass–spring–damper' molecular systems. The polarisation of a molecule measures how far an electric field at a particular frequency stretches the molecular structure, so within α_j summed over all resonances we have the entire behaviour of the material's dielectric response to an electromagnetic field – in other words the variation of its dielectric constant with frequency.

The dielectric constant is related to the refractive index n, the principal parameter of interest in photonics, through $\varepsilon_r = n^2$, and the 'damper' component gives rise to a complex solution for the dielectric constant which will vary with frequency. The size of the 'damper' can be estimated through the linewidth of the spectral absorption line. In the photonic region most of the resonances concerned are associated with electronic transitions but many also have mixing effects with molecular vibrations, exemplified in the line spectra of gases, as illustrated in Figure A3.2. Spectra are also modified through changes in pressure, introducing more molecular collisions and therefore broadening the spectral line until individual lines are completely lost, as in the spectra of liquids and gases.

The damper element represents losses in the molecule either due to the absorption of light and its release as heat or due to collisions and the transfer of energy among different molecules. This loss term operates analogously to that in all resonant systems, exemplified in the electronic LRC circuit, and is reflected in the resonance linewidth. In some cases the loss term can be safely neglected, for example, dielectric effects in metals above the plasma resonance.

Figure 2.7 shows the situation in a material with free electrons, each of effective mass m_e, with electronic charge q_e and with free electron density N_o. The plasma frequency ω_p is given by

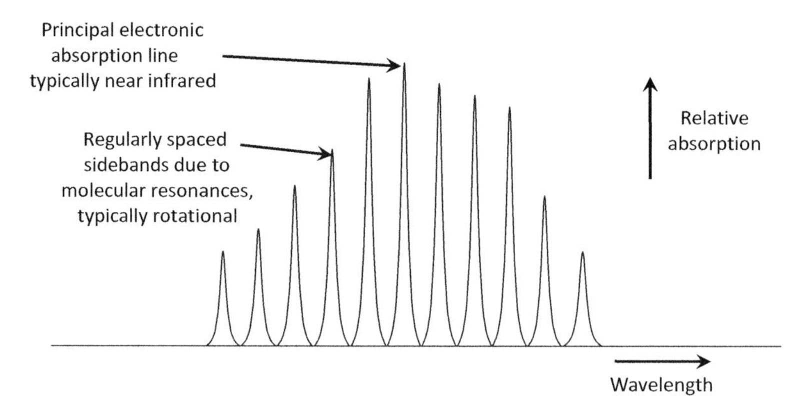

Figure A3.2 A simplified line structure for absorption in gases around the near infrared (around 1 to 3 microns wavelength). There is a central peak with numerous rotational sidebands, typically about 2 to 3 nm apart.

$$\omega_p{}^2 = q_e{}^2 N_0/(m_e \varepsilon_0)$$

Here ε_0 is the permittivity of free space. This relationship can be deduced by considering the motion of a solid block of electrons, conveniently in a cube or sphere, for which the total charge is at the centre of the cube or sphere but is displaced from the equivalent block of stationary ions (with an equal and opposite charge at its centre). The block of electrons experiences an electrostatic restoring force. It then simply equivalent to a mass on a spring. This simple approach to finding the resonant frequency ω_p gives the same answer as a more general approach which considers the charges to be in random locations!

This account has moved quickly through the basic ideas without lingering on detailed mathematical proofs, all of which are available in a multitude of textbooks or on-line scientific sites. On a more general front, what we have outlined is but one example of the wide ranging insights to be gained from the simple mass–spring–damper oscillator whether in acoustics, in the waving of trees in the breeze, in the vibrations of car suspensions or even in atomic and molecular spectroscopy.

Further Reading

J. H. Hannay 'The Clausius–Mossotti equation: an alternative derivation', *European Journal of Physics*, Volume 4, No. 3, p. 141, 1993.

C. Kittel, *Introduction to Solid State Physics*, 8th edition. First published in 1953, Wiley, 2008.

A. Kumar, *Introduction to Solid State Physics*, PHI Learning, 2010.

Appendix 4
The Concepts of Phase and Group Velocity

The ideas of phase and group velocity for a travelling wave have an important role to play in photonics, and also in any other wave related transmission phenomena. A whole chapter in Pierce's splendid text (see the Further Reading section) is dedicated to the topic. The phase velocity, v_p, is the speed with which a wavefront at a particular frequency travels through a medium. The group velocity, v_g, is the speed with which a wave packet containing a spread of frequencies travels though the same medium. Implicitly this spread of frequencies is an essential feature if the wave is to carry energy or information. Another essential feature is that these ideas are most apparent in material structures rather than in a uniform material space without dispersion: in such a space, the phase and group velocities are the same. Remember though that here the term 'structure' can include variations in material properties over the spread of frequencies in the wave packet, so for example when there are material resonances the differences between phase and group velocity will manifest themselves. Another important implication is that, whilst the speed with which energy travels cannot exceed the speed of light in the medium, the speed with which the wavefronts move can exceed this velocity. Recall that the refractive index refers to a particular frequency, hence to the phase velocity; all that's travelling here is an important abstract concept!

Here, we will briefly explore the meaning of phase and group velocity and their representation in what has become known as the omega–beta (ω–β) diagram, an example of which is shown in Figure A4.1. This figure indicates a structure which does not transmit below the cut-off frequency shown and for which the phase velocity consistency exceeds the velocity of light c in the transmitting material, going to infinity at frequencies below the cut off. As a comparison the dispersion diagram for a very large section of this unstructured material is also shown. For this baseline material, there is zero dispersion (no velocity variation with frequency) and so both phase and group velocities are c. In practice, this dispersion curve will only strictly apply to a vacuum.

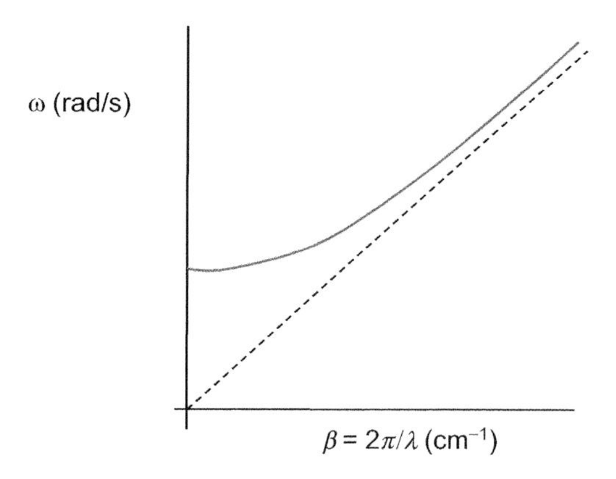

Figure A4.1 A typical ω–β dispersion diagram for a structured system, with that of a corresponding non-dispersive uniform material shown as a dotted line.

We have seen the parameter ω, often referred to as the *angular frequency*, seen many times before – it is $2\pi f$ where f is the frequency of the wave. The parameter β, referred to as the *wave number*, is less familiar and is defined as $2\pi/\lambda$. The ratio of these two quantities, ω/β, is often referred to as the *wave velocity*. It is more correctly termed the *phase velocity*, v_p, of the wave. This is the velocity of a specific phase front at the frequency concerned.

The velocity at which energy flows at this frequency is in general different, as mentioned above. In order to transmit energy, there is by definition an implied modulation of the basic sinusoidal wave, since the wave train has to be of finite extent; consequently the wave train is no longer at a single frequency: there is a finite bandwidth, determined by the modulation function implicit in the transmission of energy or data. In this case the flow of energy carried by the total waveform travels at a velocity which is generally different and which is defined as the slope of the ω–β diagram, $d\omega/d\beta$.

As an example, we will consider the case of an electromagnetic wave travelling between two perfectly conducting parallel plates, to illustrate how the phase and group velocity can be very different (Figure A4.2). Whilst this is somewhat contrived, it is conceptually close to many waveguiding principles and provides a useful insight. It is actually a special case of a general rule – the essential concepts apply to any structure through which a wave can propagate.

The figure shows two interfering broad equal-amplitude parallel light beams with the wavefronts indicated. The wave front lines are one wavelength apart, but the solid lines and the dotted lines represent the peaks and troughs of the wavefront. Consequently, when these wave fronts interfere there will be a zero in the interference pattern. The condition shown corresponds to a zero at the

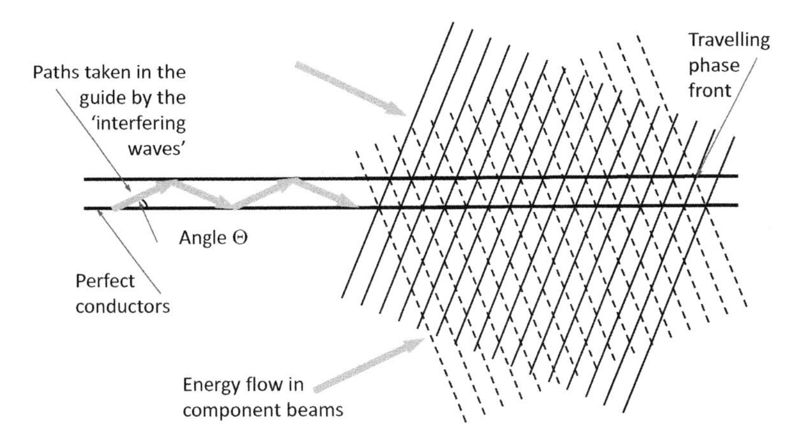

Figure A4.2 A conceptual diagram of the beam components of a mode in a parallel plate waveguide surrounded by perfectly conducting walls, and the origins of the dispersion phenomena therein.

interface with the metallic guide surfaces, with no additional zeroes in the interference pattern between the walls.

The energy in these component wave fronts travels in the directions indicated and at a velocity c dictated by the medium, which we take as a vacuum. However, the energy propagating along the waveguide will flow at the component of this velocity along the waveguide – in other words at $c \cos \theta$. This is the group velocity along the guide. Note also that the precise angle required in the interfering beams to produce the necessary patterns will vary with frequency. Hence we shall see group velocity dispersion.

What about the phase fronts travelling along the guide? The points at which the indicated interfering components cross at the waveguide-to-air- interface are the travelling wavefronts along the guide. If we take the wavelength in vacuum along the direction of travel of the component waves as λ then the distance between these crossing points along the direction of the guide – the phase wavelength – is $\lambda/\cos \theta$. The phase velocity is then the phase wavelength times the optical frequency – namely $c/\cos \theta$. You might have noticed that the product of the phase and group velocities in this case is c^2. This is a general rule, with only a very few, rarely encountered, exceptions.

Further Reading

J. R. Pierce, *Almost All About Waves*, Dover Books, 2006. Reprinted from the original 1974 text.

Appendix 5
Some Fundamental Constants

Planck's constant $h = 6.62 \times 10^{-34}$ J s.

Energy of a photon = hf, where f is the optical frequency in Hz.

Optical power, P, is related to photon energy through $P = Nhf$, where N is the rate of arrival in photons per second.

Sometimes you will see $\omega = 2\pi f$ referred to as the frequency. Technically it should be the angular frequency, in radians per second.

Boltzmann's constant $k_B = 1.38 \times 10^{-23}$ J K^{-1}.

Permittivity of free space $\varepsilon_0 = 8.85 \times 10^{-12}$ F m^{-1}.

Permeability of free space $\mu_0 = 4\pi \times 10^{-7}$ H m^{-1}.

Mass of the electron $m_e = 9.1 \times 10^{-31}$ kg.

Electronic charge $q = |e| = 1.6 \times 10^{-19}$ C.

Avogadro's number $N_0 = 6.02 \times 10^{26}$ atoms per kg mole.

The electron volt, the energy change of 1 electronic charge moving through 1 volt potential, 1 eV $\equiv 1.6 \times 10^{-19}$ J.

Photon energy = hf joules ~ $1.24/\lambda$ eV for a wavelength measured in microns.

Dielectric constant or relative permittivity for silicon, ε (Si) = 11.9.

Velocity of light in a vacuum, $c = 3 \times 10^8$ m s^{-1} = $1/(\mu_0\varepsilon_0)^{0.5}$.

Room temperature can be taken as approximately $T = 300$ K.

Appendix 6
Comments and Hints on Chapter Problems

To start with, we give a quick recap on the comments in the Preface. To reach solutions, there will often be a need to look up sources and to appreciate trade-offs. Central to solving problems is the idea that you discuss any approach with others: we can certainly learn from our fellows in addition to our formal teachers. Often, there is no such thing as a totally right answer, especially to any real engineering problem. So, collectively, work out the best approach and go for the best solution you can find. Then, very importantly, critically analyse this solution. Make sure it stands up to scrutiny! Do not forget, that there is much source material on the web but, even here, critically assess what you find. Very occasionally it can be misleading.

Also, exchange your views through the community; there will be web-based forums from which you can also learn from the experiences of others.

A6.1 Chapter 1

The principal theme of all four of these Chapter 1 problems is the following question. 'When do photons really matter in our endeavours to usefully understand the world around us?' Associated with this question is the matter of the sometimes erroneous or ambiguous use of scientific language.

Problem 1

This concerns the ideas of creating and detecting electromagnetic radiation. It is reasonable to work with the model according to which transmission always takes place as electromagnetic waves. However, the way to view the creation and detection processes varies depending on circumstances. The temperature in deep space can be rapidly determined by a question in your web browser. It can be safely approximated as 3 K, around 1% of room temperature. Thermal particle energies, which manifest themselves as thermal noise, are then around 250 μeV. Taking the photon energy as about $1.24/\lambda$, with λ in microns, gives the wavelength of a photon with this thermal energy photon is around 5 mm (you can also scale this from Figure 1.2). So photons will come in at about 3 mm and electronics at about 3 cm; again this can be scaled from Figure 1.2.

Problem 2

Microwave photonics is an ambiguous term: it is used occasionally in situations when microwaves can be viewed as photons (and you may dig around on the web for suitable examples). More often, though, it refers to the applications and technologies around modulating an optical signal at microwave frequencies (typically in the 1 to 100 GHz band). Optical communication applications in deep space are beginning to look promising – so is one photon modulating another? Is this a combination of electromagnetics and electronics? This is a fruitful topic for discussion!

Problem 3

This is a straightforward extension of the thought processes above . . .

Problem 4

This highlights the much vaunted wave–particle duality discourse. Look up X–ray diffraction, then think of the implications – and how hot would the world need to be for X-rays to be normally viewed as being propagated as electromagnetic waves in a medium? This estimate also illustrates the pragmatic approach to waves, particles and electric currents explored in Chapter 2.

A6.2 Chapter 2

For the remaining chapters, we will discuss some of the problems, maybe a sample or maybe all of them, and then encourage you to explore answers among yourselves.

In this chapter, perhaps problem 4 is the most pertinent. In problem 4(a) – deriving Snell's law – the key here is that the projections of the wavefronts onto the interface must match on each side and also that the phase velocity of the incident and refracted rays on each side is determined by the relevant refractive index. Another way of saying this is that the optical phase velocities are the same on each side of the interface between the two different materials; as mentioned before this is an important generic observation. Also, when you sketch the situation, notice that the wavelength components along the interface exceed those in the input beam itself. If the input is from air or vacuum, then the phase velocity at the interface exceeds c! This is a useful conceptual tool to visualise situations when $v_{\mathrm{phase}} > c$.

For problem 4(b), look around and think about everyway phenomena – and these can be acoustic or mechanical as well as visual.

In Problem 4(c), full derivations for the Fresnel reflection coefficients are quite involved, but the final results are very definitely of interest. In terms of *amplitude* reflection coefficients and transmission coefficients we have, for waves incident from a region of index n_1 entering a region of index n_2 via a planar interface;

$$t_s = 2n_1 \cos\theta_1 \,/\, (n_1 \cos\theta_1 + n_2 \cos\theta_2)$$

$$r_s = (n_1 \cos\theta_1 - n_2 \cos\theta_2) \,/\, (n_1 \cos\theta_1 + n_2 \cos\theta_2)$$

for the S polarisation, for which the electric vector is perpendicular to the plane of incidence. For the P polarisation,

$$t_p = 2n_1 \cos\theta_1 \,/\, (n_2 \cos\theta_1 + n_1 \cos\theta_2)$$

$$r_p = (n_2 \cos\theta_1 - n_1 \cos\theta_2) \,/\, (n_2 \cos\theta_1 + n_1 \cos\theta_2)$$

Note the subscripts in the P and S cases for the transmitted and reflected amplitudes...

Other points to note here:

- The phase differences depend on whether the reflection is from high to low index or vice versa. There is a 180° difference (multiply by –1)
- Check that the sum of the transmitted and reflected **power** coefficients is what you would expect it to be – namely unity. Power is proportional to the square of amplitude.
- You are looking through a glass window, with refractive index 1.5. How much light will be reflected at normal incidence (our eyes, like almost every other photodetector, respond to power)? Consider viewing an illuminated object from different angles?

A6.3 Chapter 3

The problems here all explore what happens when light passes through a material. Here, the basic possibilities are absorption and scattering, absorbed light turning into heat, light at other wavelengths, electronic carrier generation, refractive index changes. Consider what happens at interfaces, as already mentioned in Chapter 2. In some places the discussion here is a preamble to topics discussed in Chapter 4, for example, birefringence.

Problem 1

This concerns the possible ways in which radiation from the sun (assume a black body at the temperature mentioned in the text) is altered by passing through the earth's atmosphere. First, calculate the spectrum as a function of wavelength. The consider the various phenomena which can modify the spectrum between its arrival at our atmosphere and its arrival at earth's surface. You will need to explore absorption and scattering and also the ways in which these phenomena vary with weather conditions, where for example both scatter (from clouds in particular) and atmospheric content (water vapour) can change. There are also implications regarding altitude – how would things change at 8000 m?

And, as for most of our problems, there is scope for teamwork in this problems. Again, there is no totally right answer, so a good basic understanding is essential in order to weigh up the different possibilities!

Problem 2

The approach here could be to start by taking 1 mW as the same power level all along the fibre, for 50 km, and see what comes out of that. Thereafter, examine the situation where the average loss is a typical value (look this up – or assume, for a start, that it is 0.5 dB per km). What is the input power needed to average 1 mW over the fibre length? Integrate this distribution over the length to obtain the total Kerr-induced phase change. Next, examine the same process for, say 1 dB/km and 0.1 dB/km.

Part 2 of the problem is about the mixing processes which occur in non-linear media and the generation of sum and difference signals. You will need to think this process through, but the clue lies in the fact that changes induced by frequency 1 will also impact on the transmission properties of frequency 2. As for 'significant' – you can access the literature on four-wave mixing in optical communications when two signals very close to each other in wavelength are transmitted along a fibre, or you can decide on appropriate sounding criteria and then compare with actual practice. An appropriate starting point might be that the amplitude of the sum and difference signals is 10% of the separate signals (20 dB down), for example. In cases like this it is always helpful to initially establish some guidelines to get a feel for the situation of interest.

Problem 3

This is about detecting light and how detection is almost inevitably a function of optical wavelength. The differences between bandgap (photon) and bolometer (power-into-temperature-change) systems are discussed in the text. You might like to analyse this further, however, and think about the impact of the modulation bandwidth on the performance of the two types of system and also the achievable signal to noise ratios for each type – how small an input power change could be detected by the bolometer and the (shot noise limited) bandgap detector? How might these thresholds vary with the optical incident power?

The final part of this problem, concerning the design of an 'ideal' photodetector, requires some thought on what is ideal. Yes, it needs to respond over the wavelength region specified. But over what input power range is it not specified? Here, making some assumptions seems appropriate. Start with an input of 1 mW and then consider the design implications for inputs of 1 microwatt and 1 watt? Also, we have not stated the operational bandwidth for modulated optical signal. However, here it is reasonable to assume for a start that the source is continuous.

Working from these assumptions leads into a starting point based on a small thermally well insulated bolometer with appropriate thermal conductivity coefficients (look up some reasonable values) designed into a suitable resistance measurement bridge. In contrast you will have a shot noise limited bandgap detector with the implications on responsivity previously explored. Could band gap detectors be stacked to enhance the lower wavelength responsivity?

There's much to explore in this problem, and again it is a discussion topic among a group more than an individual endeavour.

Problem 4

This is about Fresnel reflection from interfaces and optimising the combination of the reflection from the lens-to-air interface together with the LED-to-glass interface with a view to getting the maximum power release (remember that we are usually looking at amplitude reflection (see the comments above on the Chapter 2 problems).

The other aspect here is that it is possible to design coatings made from a material with an intermediate index, which for one single wavelength will allow though all the light from one medium to another. There is a Wikipedia article on optical coatings, and you may wish to demonstrate the expression for the matching-layer thickness mentioned there using the Fresnel relationships. So, in principle, for a single-wavelength LED, with careful design all the light could be emitted.

As for the laser issue: coatings are an important factor here, but this time they are used to reflect, rather than match for transmission. The basic aim with this question is to encourage an understanding of the lasing process, the thinking that eventually leads to the laser equation and the concept of recombination rates. Figure A6.1 may provide a useful starting point.

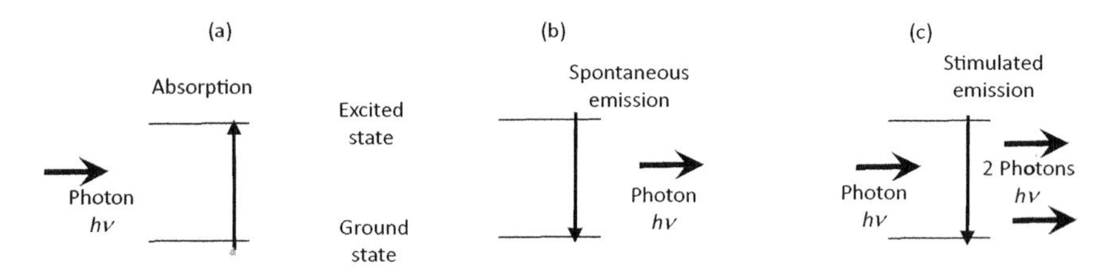

Figure A6.1 A representation of the three processes possible in photon generation and absorption. (a) Straightforward absorption of an input photon which results in thermal generation or sometimes the generation of a lower energy photon via fluorescence for example. (b) represents spontaneous emission where an excited state, for example produced by electrical current injection, produces a photon which is not correlated with other photons produced by the input excitation. (c) - stimulated emission, the basis of optical amplification and lasing, where an input photon causes the emission another one of the same wavelength in an already excited material.

Problem 5

This problem is about the way in which plastics induce stress-related birefringence between two crossed polarisers, so that a component comes through the second polariser. In general this component will be coloured, since the birefringence will change (almost double) in going from long (red) to shorter wavelengths.

With careful analysis this can be quantified. However, accurate knowledge of the stress-optic coefficients involved is sparse (but how would they be used?), and in practice a very rough guide can be derived from the variations in colour. Blue is more sensitive (less stress needed to introduce this colour) than red. Why is this the case?

Problem 6

Here are some hints on this problem.

- In depletion layer an applied voltage will remove all the free carriers. This leaves behind the charges on the remaining lattice at a charge density equivalent to the doping (free-carrier) density times the electronic charge. This can then be put into Gauss's law to obtain a field slope. However, for an intrinsic region it is a reasonable approximation to assume that remaining charge density is low enough that the slope of the field is zero in this region. In the p^+ and n^+ (substrate) regions it is also safe to assume zero field slopes. This simplification gives the need for a bias voltage of 20 kV/cm times 10 microns to achieve saturated velocity. The calculated bias voltage is worthy of comment!
- The depletion layer width has to be traversed in a time which is small compared with any modulation on the optical signal. You may like to consider why this is the case, and also what might happen if, for example, the modulation frequency is twice the inverse of the transit time. The carrier velocity can be taken as 10^7 cm/s.
- The next issue lies in how far the photons will reach into the depletion region before the carrier generation rate drops below, say, 10% of the maximum reached very close to the surface. The p^+ contact is 1 micron thick. Can this thickness be ignored? How significant is this absorption of photons in the overall quantum efficiency of the device? How do all these answers change with wavelength at photon energies above the band gap? The graph at the end of this section shows carrier velocity plots in silicon and gallium arsenide (see Figure A6.2).

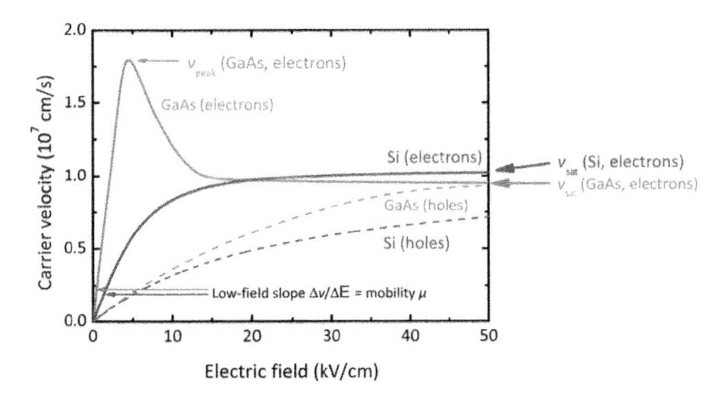

Figure A6.2 Carrier velocity vs Electric Field Characteristics for Si and GaAs.

– This discussion should have indicated that, while the basic design of a photodetector is relatively straightforward, the proverbial devil lies in the detail. In particular, compatibility with external circuit values (bias voltages and dimensions especially) and operational bandwidth come to the fore, coupled strongly with the operating wavelength. These considerations form another topic for discussion!

A6.4 Chapter 4

Problem 1

For a thin lens, perhaps the simplest way to conceive this might be to consider a stack of thin isosceles-triangle prisms made from the lens glass, and then to calculate the angle of exit from this prism as a function of distance from the principal axis of the lens. The angles in the isosceles triangle will be determined by the slope of the curved surface of the lens at a particular point.

To get the perfect curvature for focussing regardless of the distance from the axis, i.e. away from the paraxial approximation for a thin lens, all that is needed is to generalise the paraxial approach used above – but to use it 'backwards' in order to obtain the necessary slope as a function of distance from the axis.

This will not give the exact answer or a prescription for a practical lens, but it will give the general idea. And it is interesting to discuss the reasons why it won't be exact.

Problem 2

This is about a generalised approach to Snell's law, but in a variety of possible planes of incidence. This would be a good problem for group discussion, using a large sheet of paper on which to sketch the various possible situations.

Problem 3

The first part concerns Rayleigh scattering, the route through the atmosphere and the inverse fourth-power law. But the story for clouds will need you to explore a somewhat different type of scattering – Mie scattering.

Problem 4

The first part is about the geometry of a parabola and 'specular' reflection, for which the angle of incidence equals the angle of reflection. The second part is about image receptor sizes, and what to do to compensate for the many sources of turbulence. When one considers the logistics of realising these ideas, things really become interesting.

Problem 5

This is simply diffraction – with a laser pointer and a piece of cloth. Quantify the process, measure distances, find the wavelength for the pointer (or make an informed estimate) and explain the multiple spots which appear in terms of a Fourier transform which could describe the essential features of the cloth. And try a different sample – a very sharp edge for example.

Problem 6

This is basically a manipulation to find the Fourier series for $A + B$ cos kz, for $A < B$ and even for $A << B$. Thereafter there are hints in the main text.

Problem 7

The first part of this question is about integrating all the possible diffraction contributors at an 'input angle = output angle' condition (see Figure 4.12). Incidentally, why should these two angles be equal? Does the idea work for unequal angles? The integration process is somewhat involved though straightforward.

As for the second part, the answer is that the frequency of the directed beam is shifted by an amount equal to the ultrasonic frequency in the travelling wave. The wavelength of the ultrasound in water (sound velocity ~1500m/s) needs to be small compared with the optical beam width and also small enough to produce a diffracted beam well separated from the input. The normal Doppler-shift concepts apply, but they do need to be carefully configured to fully appreciate what's happening here!

Problem 8

The first part implicitly points to an assumption that the electric fields in the waveguide are zero at the interface between the high-index (core) and low-index (cladding) materials. You will then need to look at a fixed guide width and construct angles at which these conditions can be met using two interfering beams directed symmetrically about the guide axis. There will be a set of discrete angles for the emerging beams, and there will also be a threshold at which these angles will correspond to incident beams at angles exceeding the critical angle between core and cladding. Then the cut-off for that waveguide has been reached.

The second part asks you to critically analyse these results and deduce why, and by roughly how much, these first intuitive estimates need to be modified to get closer to actuality. This is another topic for exploration and discussion within your group.

A6.5 Chapter 5

Problem 1

This problem concerns the design for an optical receiver. First you will need to calculate the shot noise (along the lines of the calculations underlying Figure 5.1 but for a different wavelength). Then there is the question of how much shot noise there is in the specified bandwidth. However, is it really necessary to consider this aspect at this stage? In fact all that is needed is the minimum load resistance needed for the shot noise to exceed the thermal noise.

This defines the resistance needed for the load, which should be as low as possible to get the necessary RC, i.e. load resistance times local capacitance (the latter assumed here to be the photodiode operational capacitance). Is this value of capacitance, needed for a 1 GHz bandwidth, compatible with a viable photodiode? If not, then what can be done about it? Lots of scope here for critical insights and productive discussion.

Problem 2

This requires a combination of appropriate SNR (signal to noise ratio) estimates for shot and thermal noise, but in the context of a photodiode array. You will also need to make assumptions about the total power from the illuminating source and the spectral uniformity of this white light source.

The other aspects of this problem include consideration of the aperture of the lens used to focus the absorption spectrum onto the array and the resulting 'diffraction-limited blur' on the spot on each detector. Is it OK to assume that the lens will be diffraction limited, and if not what would you do about it? The final point is to highlight the immense potential contributions from the use of long integration times, reducing the bandwidths to a few Hz or often far less, in the detection process. Looked at in reverse, there are many atomic processes which can be best understood using transient (nanosecond and better) observation. What are the implications here?

Problem 3

The whole of this story can be deduced from Figure 5.3, the text around it and some basics on Fourier transforms: notably, what determines the resolution and dynamic range in a Fourier transform calculation? The only distraction is that the resolution in the first part of the question is specified in wavelengths, whereas the Fourier transform system fundamentally works in optical frequencies. Additionally, the frequency range is in wavelength terms (sometimes you will see spectroscopic results in terms of 'wavenumbers' – the inverse of wavelengths).

Problem 4

Part (a) of this problem is in principle straightforward – what is the minimum bandwidth on a 1 micron source required to achieve a coherence length of less than 5 microns? There is some juggling to be done with the units (time, distance, frequency, wavelength in microns and bandwidths in nm) used here. There is also a very important but unstated need here in that the source needs to be spatially coherent. Another unstated need is depth resolution, but in what medium? It is probably safe to assume that this medium is air, since all the other media will have a refractive index greater than or equal to 1. Why is this the case?

The second part of the problem brings together the mechanics and drives, the mirrors, the data storage and processing for the whole system and the need to put numerical requirements on them all. Hence explore among your group how practical it might be to go to 1 micron, or even better, spatial resolution.

The last part of the problem is about illumination wavelengths, the transparency of the sample and the wavelengths which might characterise artefacts of interest in a particular sample. There is some interesting and useful exploration to be done around these issues.

Problem 5

The first part of this question is basically a standard derivation, but the ideas should be within the grasp of readers without any need for hunting elsewhere. Think about the basic situation. The light is injected in both directions at once from a particular point (the beam splitter or directional coupler). It then propagates around the loop, and one can assume for simplicity that it propagates in air. After a delay in the loop, the light then meets itself coming in the opposite direction, as it were, but the lengths of the clockwise and counter-clockwise paths will be slightly different by an amount corresponding to the rotation rate times the time delay (multiplied by 2 or not? Think that one through carefully).

This time delay corresponds to a phase delay and the rest is straightforward. Interestingly, the result is independent of the refractive index of the medium through which the light is travelling around the loop.

The second part concerns making reasonable estimates. How long could the fibre be without too much loss, allowing for consistent bending? How sensitive could such a precision fibre interferometer be, when designed by a skilled person?

Problem 6

Straightforward radiation pressure calculations are needed for part (a), and some careful descriptive discussion around the situation which pertains for part (b), including the very pertinent query, in which direction does or should gravity work?

A6.6 Chapter 6

Problem 1

The key to this problem lies in the fact that lithium niobate is electro-optic: its dielectric constant can be changed by applying an electric field via the surface electrodes. Therefore, the phase delay in the upper path in the diagram can be modulated. This is an interferometer and, from the symmetry in the design, it can be assumed that equal optical power goes along each path. Consequently, if the two paths emerge via the 'light out' path in phase, all the optical power goes into that port. (The two paths will be in antiphase at the other port, so nothing exits.) Consequently, electrically changing the phase difference changes the output. There are more subtleties in this when the output ports are not in phase.

In part (b) the frequency limits are to do with transit times through the modulator and what RC time constants are possible. What would be needed for (i) a phase modulation system and (ii) a frequency modulator? You may also like to consider a frequency shifter system. Although this ventures into communication system theory, it is relevant since such systems are in use in optical communications.

Finally, another aspect of this – what might happen if these functions were implemented in silicon photonics? This leads into the next problem.

Problem 2

The answers to this problem (and more background on the foundations of silicon photonics) can be gleaned from G.T. Reed *et al.* 'Silicon optical modulators' in *Nature Photonics*, vol. 4, pp. 518–526, August 2010, and other articles in the 'Focus' special section on silicon photonics in this issue.

Problem 3

The first part of this problem is about phase matching along the interface. You will need to dig a little to derive an accurate value for the phase velocity of the wave along the interface. However, some useful insight into how this all works is gained by making some 'first guesses'.

Note that, whilst this is rarely stated, the top overlay (often referred to as the 'analyte') is often water based and is implicitly of a lower refractive index than the substrate material. The phase velocity along the interface is consequently higher than in the bulk material. We also have a mirror in the metallic layer and this will preferentially reflect the incident wave, unless there is an exact coupling into this hybrid 'substrate–metal–analyte-layer' wave structure (as indicated, and with an effective index close to the average value between the two sides of the very thin metal layer). At this point there will be a dip in the reflected wave when this coupling takes place, corresponding to phase matching between the substrate incident wave and the surface wave defined by the properties of the interfaces of the analyte, thin metallic and substrate layers.

There's a nice review (30 pages) of the principles of all this in K. M. Mayer and J. H. Hafner, 'Localised Surface Plasmon Resonance Sensors', *Chemical Reviews* 2011 (dx.doi.org/10.1021/cr100313v | Chem. Rev. 2011, 111, 3828–3857).

The second part of this problem on TM or TE polarisation states at the interface was discussed in the text. As for the 'needle' structure in (b) in the figure: think about the incoming direction of the incident wave, the dimensions of the 'needles' and the operation of a lightning conductor!

Problem 4

The basic feature here is that, by changing the bias on the central electrode, it is possible to change the delay time for the current flowing through the graphene by the equivalent of 180° at 10.6 microns, which corresponds to about 28 THz or a period of about 35 fs. Thus the time-delay change needed across the ~300 nm device corresponding to this change in transit time is equivalent to a change in the electron velocity of ~20×10^6 m/s! This should be checked carefully. It gives some insight into the velocity of the carriers in graphene compared with the velocity of the carriers in silicon (see Figure A6.2).

You could also work through the sheet resistance figures in Figure 6.12 to gain some insight into the mobility of electrons as carriers in graphene. Investigation into the electron density in the graphene sheet is needed in order to gain this insight. You will need investigate the basic ideas of sheet resistance.

Problem 5

For this problem, look at the order of magnitude of the dimensions of the resonator as shown in the diagram. Might the structure have a resonance in the optical range? The overall device does need to be cooled down to work; thermally induced vibrations in such a tiny structure could cause considerable trouble. They would also reduce the effective optical Q factor of the cavity (why?). Even so, only 0.5 dB of squeezing is produced.

The original idea was published in A.H. Safavi-Naeini *et al.*, *Nature*, Vol. 500, p.185, August 2013. There was also a follow up based on silicon but with a somewhat better performance, this time using very high Q factor resonators. You can find this at arXiv1309_6371v1 where it is openly available.

As for the future – well – what do *you* think?

Problems 6–8

These problems are intended as joint research projects with your colleagues. They give an opportunity to explore factors which will come into play as photonics continues to expand both its contributions into society and its 'tool kit' of esoteric devices performing intriguing, in many cases as yet undreamt of, tasks. There is a fascinating future yet to evolve – we are just at the beginning!

Further Reading

A. On Background Tools and Concepts

1. J. R. Pierce, *Almost All about Waves*, Dover Publications, originally 1974, 200 pp. Remains available.
2. R. Bracewell, *The Fourier Transform and its Applications*, McGraw Hill, originally 1965, 375 pp. Remains available.
3. J. F. James, *A Student's Guide to Fourier Transforms*, Cambridge University Press, 2002, 130 pp.
4. D. Fleisch, *A Student's Guide to Maxwell's Equations*, Cambridge University Press, 2008, 130 pp.
5. E. Collett, *Field Guide to Polarization*, SPIE Field Guides, Vol. FG05, SPIE, 2005.
6. J. W. Goodman, *Introduction to Fourier Optics*, first published 1968, 4th edition, McGraw Hill, 2017, Remains the most respected text.
7. R. Feynman, R. Leighton and M. Sands, *The Feynman Lectures on Physics*. Three volumes, 1964, 1966. Library of Congress Catalog Card No. 63-20717 nISBN 0-8053-9045-6 (2006 the definitive edition, 2nd printing, hardcover). Well presented, well argued, accounts of subtle concepts in optics and photonics (and most of the rest of essential physics too!).
8. The 'Optics for Kids' web site (https://www.optics4kids.org/home) run by the Optical Society of America (OSA) also has some interesting accounts of many basic ideas in optics – can be fun to look through!

B. Publications Triggered by the International Year of Light

1. M. Garcia-Matos and L Torner, *The Wonders of Light*, Cambridge University Press, 2015.
2. K. Arcand and M. Watzke, *Light – The Visible Spectrum and Beyond*, Black Dog, 2015.
3. *Celebrating Light*, SPIE Press, 2015.
4. *Inspired by Light*, SPIE Press, 2016.

C. More Details on Photonics Principles

1. G. R. Fowles, *Introduction to Modern Optics*, Dover Books, 1990.
2. B. Saleh and M. Teich, *Fundamentals of Photonics*, Wiley, 2007, 1175, pp.
3. J. Donnelly and N. Massa, *Light – Introduction to Optics and Photonics*, New England Board of Higher Education, Boston, 2007.
4. M. Mansuripur, *Classical Optics and its Applications*, Cambridge University Press, 2009.
5. E. Hecht, *Optics*, 5th edition, Pearson Press. This book first appeared in 1974 and continues as an authoritative and comprehensible guide to wave and ray optics.
6. D. Voetz, *Computational Fourier Optics*, SPIE Tutorial Texts, Vol. TT89,
7. G. O. Reynolds *et al.*, *Tutorials in Fourier Optics*, SPIE Press, Bellingham, WA, 1989.
8. There is also a free IPad App which performs spatial Fourier transforms on photographs stored on the device – very instructive!

D. Some Experiments

1. R. N. Zare, B. H. Spencer and M. P. Jacobson, *Laser Experiments for Beginners*, University Science Books, Sausalito, CA, 1995.
2. R. N. Compton and M. A. Duncan, *Laser Experiments for Chemistry and Physics*, Oxford University Press, 2016.

E. Some Examples of Online Resources on Photonic Principles and Applications

1. http://spie.org/?&ID=x10&SSO=1 for free texts including educational CDs and videos.
2. http://spie.org/publications/fundamentals-of-photonics-modules for photonics principles modules.
3. http://spie.org/education/education-outreach-resources/online-resources lists a rich variety of educational material available free from SPIE.

F. The Troubles About Waves, Rays and Photons

1. F. Sellen, *Wave Particle Duality*, A compilation from numerous authors totalling 324 pages, originally published in 1992 by Springer; published in paperback, 2013.
2. R. Chandrasekar *Causal Physics: Photons by Non-Interactions of Waves*, CRC press, 2014. A single-author perspective targeted towards optical science and engineering.

G. Some More Advanced Resources

1. A. Cutolo, A. G. Mignani and A. Tajani (eds.), *Photonics for Safety and Security*, World Scientific, 2014. An excellent compilation of photonic technologies and applications.
2. SPIE Tutorial texts in Optical Engineering – a series of over 100 straightforward tutorial texts in topics from spectroscopy to nanophotonics to plasmonics to optical design. See also SPIE *Spotlights* and SPIE *Field Guides*. These books cover all the photonic principles and applications and are invariably presented in a concise and readable manner.
3. J. P. Dowling, *Schrödinger's Killer App: Race to Build the World's First Quantum Computer*, CRC Press, 2003.
4. O. Solgaard, *Photonic Microsystems*, Springer, 2009.
5. M. Ohtsu and H. Hori, *Near-Field Nano-Optics: From Basic Principles to Nano-Fabrication and Nano-Photonics (Lasers, Photonics, and Electro-Optics)*, Springer, 2012.
6. S. V. Gaponenko, *Introduction to Nanophotonics*, Cambridge University Press, 2010.
7. L. Novotny and B. Hecht, *Principles of Nano-Optics*, 2nd edition, Cambridge University Press, 2012.
8. K. Sakoda and M. Van de Voorde, *Micro and Nano Photonic Technologies*, Wiley, 2017.
9. E. Diamanti, 'Quantum signals could soon span the globe', *Nature*, Vol. 549, p. 41, September 2017.
10. P. N. Prasad, *Introduction to Biophotonics*, Wiley, 2003.
11. G. Keiser, *Optical Fiber Communications*, 5th edition, Tata McGraw-Hill Education 2013.
12. J. Hecht, *City of Light*, Oxford University Press, 1999.
13. SPIE Tutorial texts in Optical Engineering – a series of over 100 straightforward tutorial texts in topics from spectroscopy to nanophotonics to plasmonics to optical design.
14. J. H. Hannay, 'The Clausius–Mossotti equation: an alternative derivation', *European Journal of Physics*, Vol. 4, No. 3, p. 141, 1993.
15. C. Kittel, *Introduction to Solid State Physics*, now in its 8th edition after first being published in 1953, Wiley.
16. A. Kumar, *Introduction to Solid State Physics*, PHI Learning, 2000.

H. Sources of Latest News and Developments

1. *Nature Photonics* often features special issues and tutorial articles, for example on plasmonics, graphene, silicon photonics and many other topics.
2. The Professional Societies, most notably SPIE and OSA, disseminate technology reviews though their member journals, *SPIE Professional* and *Optics and Photonics News*.

3. SPIE Newsroom combines scientific and commercial announcements in optics and photonics.
4. Trade Journals, available on free subscription, such as *Photonics Spectra*, *Laser Focus World*, and *Biophotonics*, regularly publish easily digested technology updates.
5. Consult web sites for Photonics 21(Europe) and the National Photonics Initiative (US), and also download their reports, which present technology and applications developments important for the future of society and aimed at a general, technology-literate, audience.

Index